WE ARE
AGORA

Also by Byron Reese

Stories, Dice, and Rocks That Think
Wasted
The Fourth Age
Infinite Progress

WE ARE AGORA

How Humanity Functions as a Single Superorganism That Shapes Our World and Our Future

BYRON REESE

BenBella Books, Inc.
Dallas, TX

BenBella Books, Inc.
10440 N. Central Expressway
Suite 800
Dallas, TX 75231
benbellabooks.com
Send feedback to feedback@benbellabooks.com

BenBella is a federally registered trademark.

Printed in the United States of America
10 9 8 7 6 5 4 3 2 1

Library of Congress Control Number: 2023026946
ISBN (hardcover) 9781637744215
ISBN (electronic) 9781637744222

Editing by Alexa Stevenson
Copyediting by Scott Calamar
Proofreading by Cape Cod Compositors and Marissa Wold Uhrina
Indexing by WordCo Indexing Services
Text design and composition by PerfecType
Cover design by Pete Garceau
Cover image © iStock / DigitalStorm
Printed by Lake Book Manufacturing

To Agora

CONTENTS

SECTION II
FORCES

SECTION III
AGORA

SECTION IV
MEANING

INTRODUCTION

A SUPERORGANISM IS A single life-form made up of other living organisms.

It would be easy to gloss over that short sentence of a dozen words, but it's a big idea—as big as they come. Why? Just think about yourself for a moment. You are made up of cells, that is, living creatures. But you are a different creature entirely from your cells. Isn't that strange? After all, if in some alternate universe people *weren't* made of living cells but instead composed of equally talented pieces of inorganic matter, you could still picture a regular human being—just one made out of inorganic parts. Likewise, consider the opposite: If there were a universe where your cells were living creatures but there was no "you" that arose from them, that would make sense, too, right? You basically would be a mere bag of cells, something akin to an aquarium full of fish.

But now imagine a weird, crazy universe where both of those things are true. Your cells are alive and live full, complete cell lives, and somehow, superimposed on that, is an entirely different creature; that is, you. That would be a strange reality, and, as luck would have it, that is the one we find ourselves in.

We somehow live in a universe where multiple life-forms can share the same matter. That's just mind-boggling to think about. They don't share the same body, as in one of them gets half and the other one gets the other half; they actually coexist on the same exact material. You live in your body, and your cells live in your body, but you aren't roommates. You actually share the same matter, superimposed upon itself. If we weren't so accustomed to this idea by virtue of being taught it all our lives, it would sound preposterous.

It's hard to even come up with a corollary to help conceptualize this, but let me try. Have you ever seen one of those posters of something like a kitten, and if you lean in and look at it really closely, you notice that it isn't made up of pixels but of thousands of miniscule photos of other kittens? Then, hypothetically, you could grab a microscope and look at those tiny photos and notice that they are made up of even smaller photos of kittens. Stepping back and looking at the big poster, the thought might occur to you that maybe it goes the other way, too, that the poster you are looking at might itself be a miniscule dot in an inconceivably larger poster of a kitten.

Now, think of people again. We seem to operate at just two levels of life: the cell and the human, the dots and the poster. But is that all there is? Just two levels? Maybe cells are composed of even smaller living creatures. We don't think they are, but there is nothing inherently impossible about that idea, right? Now go the other way: Is it conceivable that we in turn are cells in a larger creature that we are completely unaware of because it exists on such a different scale than we do?

We can observe things quite a bit like this in other species. Honeybees exist on at least three levels: there are cells and individuals, just as with us, but a group of individual bees forms a third level of being, the colony. A bee colony isn't a mere heap of bees but an emergent creature that is created through the interactions of the bees. They are its cells, and it is a superorganism. The colony, the beehive, is not metaphorically a living being; it actually is one.

The question that this book asks, and tries to answer, is whether humans, each and every one of us, are part of a larger living organism. Again, not metaphorically but literally. Are we cells in a larger creature?

The challenge in answering that question is hinted at above: How would we know if we were part of some larger creature? It sure doesn't seem like we are, but we can't take much reassurance in that. We wouldn't be able to perceive it, any more than our cells are able to perceive us. Humans occupy a pretty narrow band of reality, seeing only a small range of colors, hearing only a limited range of sounds. We aren't aware of things at the microscopic level, like bacteria, nor at the macroscopic, that is, planetary, level. Some infinitesimally small cell within you, or some planet-sized being that you are a part of, may be at a death metal concert right now, and you wouldn't have the slightest inkling. So, to repeat the question, how would we know if we were?

Science is the study of things that can be observed or measured in our material reality. We have built the modern world on a set of processes that we call the scientific method that advance our understanding of the nature of that reality. Metaphysics, on the other hand, is the study of the things that lie outside of that material reality, such as the nature of the mind, consciousness, being, causality, and, well, pretty much everything we wonder about but don't have a scientific way to investigate. The line between the two is fuzzy, and it shifts as science solves old mysteries and uncovers new ones. But it would be a mistake to say that scientific truths are the only ones we can ever know for certain. Rather, they are the only ones we can prove—a very different thing. In point of fact, the vast majority of what you know cannot be scientifically proven. The old reporter's adage, "If your mother says she loves you, check it out," is funny because you can't check it out, but you can obviously know it to be true.

This book is mostly about science—what we can observe and know about cells, life, intelligence, emergence, and how superorganisms such as beehives and anthills operate. But it is also about metaphysics, because it deals with mind, being, consciousness, and the nature of self. So it straddles that fuzzy line as it tackles the question of whether there is an actual superorganism comprising humans that I have named *Agora*.

Our pathway to answer this question begins in **Section I—Life**. We'll start with the origin of life and try to understand the cell, the basic indivisible unit of life. I say "indivisible" because cells are alive but are made of entirely nonliving things. They are elemental life. What makes them, these little containers of nonliving stuff, alive? What force animates them?

From there, we will move to more complex forms of life. The next step up is multicellular life. It isn't alive in the same manner as a cell. This distinction is seldom made, but you, along with beavers and begonias, are a different form of being from a cell. If I took you apart one cell at a time and deposited each cell in an appropriate medium, their lives would go on, but you would vanish. Where did you go? Were you ever even here? Maybe you aren't, strictly speaking, even alive, only your cells are. You seem more like a community than a life-form. But mysteriously, you have attributes that none of your cells have, such as a sense of humor. Where did those come from?

That takes us up another notch in complexity. Somewhere along this chain of being come creatures that *experience* the world, as opposed to just exist in the world. A cell may react to being poked with a needle, but did it have an experience of pain? No. It didn't experience it any more than a rock would. Likewise, you can feel the relaxing warmth of a hot bath in a way a cell can't. We call this experience of sensation "consciousness," and the big mystery is how mere matter can have an experience of anything. You are, after all, just a collection of cunningly arranged elements, the same exact ones that you can find in a cell or a rock, just arranged differently.

Our final step up in complexity is the thing we are really interested in. It turns out that some multicellular life-forms can come together and produce a new creature, with properties and abilities that none of its component life-forms themselves have. Examples of this include the social insects—the bees and ants—that come together to form living beings that we call hives and colonies. These hives and colonies are called superorganisms. And make no mistake, they are a new being. A beehive is alive. Not just the bees but the collective hive is itself a living creature, with different abilities and proclivities from those of any bee.

What we want to know is whether there is a human superorganism—a creature composed not of bees or ants or cells but of people.

Superorganisms all display emergent properties, that is, capabilities not present in any of their component parts. Where do those come from? That's what we tackle in **Section II—Forces**. We will discover that we can explain these capabilities by understanding the interactions of just six forces. They are energy, information, communication, cognition, specialization, and technology. We want to understand how those interactions bring a beehive or an ant mound to life. Our goal will be to see these rather mundane topics anew, to contemplate them in a general sense. Only by doing this can we apply them to a quite different thing: human society.

That will bring us to **Section III—Agora,** where we ask if the collective population of this planet forms a superorganism. Do the forces from Section II interact in humans the same way they do in bees and ants? If so, then there really is a superorganism of us; there really is Agora.

But if we wouldn't be able to perceive Agora even if we were part of it, how will we know for certain? While we cannot gaze upon Agora, we can discern its shape by the shadow that it casts. We will look at cities as hives, and people as specialized parts of a larger being. We will try to spot emergent capabilities and other hints that we are part of a larger whole.

From there, we will ask the even bigger questions. What does Agora want, if anything? Why did it evolve at all? What purpose does it serve?

Then we will ask the biggest question of all: Why are we here?

Usually, science is mum on all questions that begin with "why." When those come up, it quickly tries to change the subject. Its normal purview is "how" questions, and it drones on and on about those. But we are going to buck convention and tackle the "Why are we here?" question scientifically, and answer it without any appeal to mysticism. And if it really is a scientific question, then, as mentioned above, not only can we know the answer, but we can prove it as well. I know that's a tall order, and if you are incredulous, believe me, I understand.

Finally, we will come to **Section IV—Meaning**. We will ask if any of this matters in a practical day-to-day sense. Then we'll explore what Agora teaches us about why history unfolded the way it did, and see if that gives us any insights into our future as a species.

That's the journey ahead of us. Thank you for joining me on it.

SECTION I
LIFE

QUESTIONS

IMAGINE A PERSON WHOSE hobby is scaling cliff walls. Let's call her Celia. One morning Celia wakes up atop Mount Insanity, which she summited the previous day. Now she is ready to descend. As she approaches the edge, her foot knocks a small rock, the size of a golf ball, off the cliff. Celia watches it as it bounces from one surface to the next on its way down to the base of the mountain, and she decides she is going to follow that same path, just for a fun challenge. When she gets to the bottom, she spies the rock, picks it up, and pockets it for a memento.

If you were an observer watching all this happen, you would of course know that the rock is not alive but Celia is. If you were an alien watching, it would not be obvious. Maybe the rock is in charge and told Celia to follow it. How would the alien know? What is the main difference between what Celia did and what the rock did? Is it that Celia was making choices and the rock wasn't? That's unsatisfying because then life is turned from a biological phenomenon to a mental one. A computer playing chess is "making choices," but a sleeping human isn't. Would you therefore conclude that the computer is alive but the human isn't? Or, perhaps you would say that Celia was internally powered and that the rock was just acted upon by external forces. Or perhaps that the rock is not intelligent, or it has no metabolism. But, as we will see throughout this section, each of those potential solutions is rife with challenges.

9

The mystery is that while cells are universally regarded as living things, nothing inside of them is alive. Crazy, isn't it? The cell is just a bag of rocks, and everything that goes on inside that bag appears to be just plain-vanilla physics—nothing more mysterious than Celia's rock bouncing along. So what makes the cell *alive*? It sure isn't smart; it doesn't seem to make deliberate choices, or even have a will of its own, and certainly can't set goals for itself.

I recall the first *Toy Story* movie where the new toy, Buzz Lightyear, claims that he is able to fly. When pressed for a demonstration, he trips and stumbles through a series of comedic missteps that involve a toy car set, the ceiling fan, and so forth, that ultimately fling him through the air. Woody, another toy, is unimpressed: "That wasn't flying. That was falling with style." That's all that happens inside a cell: a whole lot of falling with style. I ask again: What makes that alive?

Life, as ubiquitous as it is on this planet, remains the great mystery. There is something that animates us and makes us different from the inanimate world around us, but we aren't sure what it is. This has puzzled the great thinkers through the ages and rightly so, for life is not a substance to be analyzed or a process that can be deconstructed. It's actually a . . . well, that's the problem. Even today we don't have an agreed-upon definition of life, a fact that I am sure is particularly annoying to biologists, given that the name of their field literally means "those who study life." On top of not knowing what it is, we don't know how it started or how it managed to achieve such mind-boggling complexity and variety. How can we know so little about something we all have so much firsthand experience with?

Perhaps there is simply a linguistic difficulty in coming up with a definition. The philosopher Ludwig Josef Johann Wittgenstein pointed out that there is no single attribute of games that is universal to all games. In other words, not all games have winners, or points, or even other players, and yet things called "games" really do exist and we can all easily agree that Monopoly is a game without having a precise definition of the word. Is it the same issue with life? That it is so diverse that we can't come up with the right combination of words to convey the right degree of inclusive and exclusive? No. I don't think that is the core problem.

Then maybe the problem is that there are so many varied constituencies that pose the question "What is life?" and they are really asking different things. Botanists might ask if certain thousand-year-old seeds are still alive, even though they exhibit no lifelike behavior, while an ecologist may ask if an ecosystem is alive. A doctor might ask if a brain-dead patient is still living in any meaningful sense, an epidemiologist might wonder if a tissue sample is alive, a computer scientist might ask the same question about artificial intelligence, while the folks at NASA pose the question to try to come up with an understanding of life that is inclusive of hypothetical beings we don't even know exist. And perhaps Dr. Frankenstein meant something different even from all of that when he maniacally yelled, "It's alive! It's alive!" All of this is unquestionably true, but none of it answers the core question; for if you ask any of those various constituencies what they mean by "life," they would likely start their answer with, "Well . . ."

Perhaps, then, the problem is all the edge cases. Does expanding the definition of life wide enough to let in something from the gray area between living and nonliving things, such as viruses, inadvertently let in other things we don't want to include, such as computer viruses? While this, too, is an issue, I don't think it is the reason we cannot define life. After all, "day" and "night" are real, definable concepts even if we cannot point to the exact instant when day becomes night.

I think our difficulty boils down to the simple fact that we still aren't sure what life is. We simply don't know. We have an intuition that fire and cars and ideas are not alive but that slime mold is, but when pressed to explain why we think this, things get complicated. Does life have to be biological? Does it have to be cellular? Does it have to reproduce? If so, or if not, why?

But I am getting ahead of myself. Let's back way up and look at the basic unit of life, the cell, and try to understand what it is and why it is regarded as living. We can then proceed to build atop that foundation.

THE CELL

THROUGHOUT THE 1600S, ADVANCES in technology related to glass and optics opened humanity's eyes to a larger view of the universe—and a smaller one as well. The two-thousand-year-old glassmaking industry in Venice, which guarded its secrets so jealously that craftsmen were forbidden to leave the city on pain of death, had a series of breakthroughs that allowed it to economically produce glass of amazing clarity that could be ground into high-quality lenses. Industrious inventors wasted little time in figuring out that if you put two of those lenses in a tube along with a mechanism for adjusting their distance from each other, you could point it at the sky and clearly see the moons of Jupiter, the rings of Saturn, and the canals of Mars. They also discovered that if they tweaked that device a bit and pointed it at a drop of rainwater, worlds even more alien would be revealed to them.

Exactly who invented the telescope or made the first compound microscope are matters of disagreement that needn't distract us. What does matter is that throughout that century, our knowledge of these larger and smaller worlds exploded. We owe much of that progress to the Venetians and their glass, but even more to a new generation of scientists—although that word was still two centuries away from being coined—who acutely felt the limits of their meager technological tool set and were obsessed with finding new ways to see and measure the world.

Glass is pretty strange stuff. It is solid as a rock, but for some mysterious reason you can see right through it. Who OK'd that? It is often an unsung hero in the story of progress. In our age it is so inexpensive and ubiquitous that it is a disposable item, but in the seventeenth century it was a game changer. It wasn't just telescopes and microscopes; glass was fashioned into the mirrors and prisms used for other scientific endeavors, as well as the heat-resistant beakers, test tubes, petri dishes, and other accoutrements of the laboratory. And, perhaps most significantly of all, it gave us the eyeglasses that added decades to the professional life of countless (ahem) farsighted Enlightenment thinkers.

The fact that glassmaking reached its zenith in Europe is one of the many interesting flukes of history. The technologically advanced Chinese had perfected porcelain—considered a superior substance—to such a degree that they had little interest in mundane glass. Chinese porcelain was so highly regarded that the Venetians made a kind of glass designed to resemble it in order to compete for the European demand of the more desired exotic import.

As the 1600s ran their course, innovation in telescopes and microscopes came steadily. In 1665, the English scientist Robert Hooke published a book called *Micrographia*, which had large, incredibly detailed illustrations of the most minute of objects, giving people a glimpse of the world beyond the range of human perception. How electrifying that must have been. Just seeing the hairs on the leg of a flea would have been fascinating on its own, but then to view objects so small that no one even knew they existed—well, that had to rock their world.

It was Hooke who pointed his microscope at a piece of cork and noted that it was made up of something he described as cells, recalling the small rooms of that name in which ascetic monks prayed. Soon after Hooke's contributions, the Dutch scientist Antonie van Leeuwenhoek made massive improvements to the microscope—which to this day we have no idea how he achieved—and glimpsed an even smaller world, that of protozoa and bacteria. He went on to calculate that there were eight million protozoa in a drop of water, a number he described as "inconceivable," which I like to imagine him saying in the voice of that guy from *The Princess Bride*. Leeuwenhoek would go on to take pretty much every substance his body was capable of producing—I'll leave out the specifics—and look at them under the new device. He was always amazed at the life he observed in it, as were the many scientists of the day who closely followed his work.

What we now call cell theory came along quickly. Within 150 years after the discovery of the cell, its three core tenets were locked down and universally accepted: cells are the basic unit of life; all living things are made of cells; and cells come from other cells. Since then, we have added three more elements to cell theory: DNA is exchanged between cells during division; organisms of similar species have similar cells; and energy flows within cells.

Before I started writing this book, my mental image of a cell was pretty simplistic, consisting of little more information than would be found on a poster hanging in a high school biology class. As you may recall, those posters generally depicted a blob-shaped cell with a few of its interior components labeled, such as the nucleus, the chromosomes, the mitochondria, and the cell's membrane. The two-dimensional nature of the posters inadvertently gave the incorrect impression the cells were flat. Finally, the drawings always left lots of empty space in the cell, presumably so that everything could swim around without bumping into each another. All of this considered, the cell seemed pretty basic—low tech, even—and one might marvel that creatures as complex as humans are made of such humble stuff.

My intuition about cells couldn't have been more wrong. The cell is staggeringly complex. Consider this: There are six million parts in a Boeing 747, and to assemble one, the workers perform fourteen thousand individual subjobs over forty-three days, following something I like to picture resembling a large set of instructions from Ikea. Now, shrink that completed 747 down so small it is no longer visible, being perhaps a thousandth the size of the tiniest piece of dust your eyes can strain to see, and you will begin to have some idea of the sheer complexity of a single cell.

The 747 analogy is not an exaggeration. The cell really does have millions of parts and is that complicated. For example, the DNA helicase—the biological motor that unwinds DNA within a cell—spins at 10,000 RPM, even faster than the engine of a 747. And complex? Every second, millions of chemical

reactions take place within a cell, all driven just by basic chemistry and physics, and its machinery is capable of making forty thousand different proteins,
all while repairing, feeding, and defending itself. And it, too, is built using a
detailed assembly manual, one written in DNA, and is built not in forty-three
days but in minutes.

The physics within the cell can be challenging to imagine. Gravity, for
instance, doesn't really work in there, at least not the way it does in our world.
A cell's trillions of atoms are grouped together as either proteins, other molecules such as water, or more complex machinery like mitochondria. If you
could somehow glimpse the inside of a cell, you would behold a world packed
so densely that you could hardly cram another thing into it, and yet it's also a
world bustling with an unfathomable amount of activity.

So what does a cell do that needs all that complexity? What's its function?
It varies, but mostly cells just follow the instructions encoded in their DNA,
much of which involves making proteins that the cell uses for various tasks.
As you probably know, almost all of your trillions of cells each carry a complete copy of the three billion base pairs of your DNA, which contain all of
the information needed to build and operate your body. That data is stored in
your DNA digitally, using a four-letter alphabet consisting of what we call *G,
T, C,* and *A*. An easy way to remember this is that the title of the 1997 movie
Gattaca, which is about genetically engineering humans, comprises just those
four letters, yet it's a made-up word never used in the film.

Your cells are packed with machines called ribosomes. In his book *The
Vital Question,* biochemist Nick Lane describes how astonishing they are. Cells
are small, of course, but ribosomes are "orders of magnitude smaller still. You
have 13 million of them in a single cell from your liver. But ribosomes are not
only incomprehensibly small; on the scale of atoms, they are massive, sophisticated superstructures. They're composed of scores of substantial subunits, moving machine parts that act with far more precision than an automated factory
line." Their job is to connect amino acids in strings we call proteins. Amino
acids are made using the GTCA alphabet of DNA. Each of them consists of a
three-letter sequence called a codon. With three letters, you can theoretically
make sixty-four different codons, but in practice your body makes only twenty.

A few of the forty-four remaining codons are used by the DNA to encode special instructions, but the rest are unused. I am skipping a lot of detail in the interest of brevity, such as how there are two additional amino acids your body can use but not code for, and of the twenty, how nine of them cannot be made "from scratch" by the body and must come from food you eat. These are the "essential amino acids."

Ribosomes take these amino acids and, following the recipe in the DNA, string them together into proteins that get folded into their individual distinctive shapes. Your average ribosome can stitch three amino acids together per second, so it can make a simple protein quite quickly but requires much more time for the longest, most complex ones, which may be made up of twenty-five thousand or more amino acids. When a ribosome's duty is done, it falls apart and its molecules are used to build whatever the cell needs next.

Broadly, though, we can group what proteins do in a few categories that give the gist of some of their uses. First, there are messenger proteins such as hormones that cells use to communicate with other cells. Next are structural proteins, such as collagen and keratin, as well as enzymatic proteins that act as catalysts to speed up certain bodily functions, such as digestion. Also, there are antibodies, which are a kind of protein that helps defend the body from illness, and finally, there are transport proteins such as hemoglobin that deliver needed materials to other parts of the body, and storage proteins such as ferratin, which holds and releases iron. Proteins also store energy, help with growth and development, and repair damaged DNA.

All of the past several paragraphs are summed up in something we call the central dogma of biology, which all life follows. It describes how information from DNA is carried by messenger RNA to the ribosomes, which use that data to build any of countless different proteins out of amino acids. Those proteins in turn run the body. One of the great wonders—and mysteries—of cell biology is how the cells know which proteins to make. Your liver cells and skin cells carry the same DNA, but they code different proteins to do their different jobs. Somehow each cell knows where it is and the position it has been drafted to play. Have you ever wondered how, even though you began as a single cell that started dividing, your body knew how to build itself with a head on one end and feet on the other? All the cells that would become your brain, your bones, your organs, and your blood began in that single cell that started dividing. How did all the new cells know what they were supposed to do? And if that weren't enough, realize that an acorn has DNA that tells it how to make an oak tree, and a sizable percentage of that acorn's DNA is identical to yours. It's a wonder that at least a few people don't grow leaves instead of hair.

In spite of all of this scientific wonder, the most unbelievable thing about DNA is actually how long it is. While it is only a few molecules wide, the DNA contained in each of your cells is six feet long. All of that is packed into a minuscule cell's even more minuscule nucleus. Yes, you read that correctly. As tiny as the cell is, somehow six feet of DNA is able to be crammed into its nucleus. And because all of your cells—except red blood cells—keep a copy of it on hand, in total, all those six-foot strands would, if stitched together, reach Pluto. That is unfathomable to me.

The border between the inside of the cell and the outside world are demarcated by an amazing structure known as a cell membrane, and all cells, by definition, have one. The membrane is not like the shell of an egg but a much more fluid, amorphous thing with materials always being added or removed to keep the cell in balance. And at the same time it has to be discerning enough to let some things into the cell and keep others out, like a savvy doorman at a hot A-lister nightclub.

How does the cell get energy to do all of this? It runs on a molecule called ATP, which powers all life on Earth. The fact that all life, all the way down to bacteria, uses it or something quite like it likely suggests that it has been around since the beginning of life. The A in it is basically the same A as in the GTCA code of your genome, while the TP is a molecule called a triphosphate. It's the third of its three phosphates that does the heavy lifting. A cell uses energy to attach that third one, then reclaims that energy by breaking it off again. So the cell can take ADP (where the DP is diphosphate, that is, just two phosphates), add some energy and a third phosphate, and have a convenient universal packet of energy. The cell sends that molecule to a place where energy is needed, and the third phosphate gets broken off, releasing that energy. How cells get the energy to put in ATP varies by cell. It may get it from food or sunlight, or even more exotic means.

You use an enormous amount of the ATP every day, something on the order of a hundred pounds of it. I know that looks like a typo, but all your cells are quite busy turning ADP into ATP, and then, when the energy is drained from the ATP and it is just lowly ADP again, repeating the process. It does it so frequently that the aggregate amount of ATP adds up to the hundred-pound figure. It isn't all that energy dense of a fuel for its weight—it is more like firewood than gasoline. So imagine if you were driving your wood-powered car across the country. Would you load it up in Los Angeles with enough wood to make it to New York? No, it would be way too heavy, so you would carry just a bit of wood and constantly get more along the way. That's what your body does with ATP—it constantly makes it, in a process that our supply-chain experts of today would call "just-in-time manufacturing." If all your cells went on strike and decided to stop turning ADP into ATP, you would die within seconds—that's how little you keep on hand.

Even with all of this, I haven't even touched on a tiny fraction of 1 percent of everything that goes on in a cell. There are so many things: how cells insulate highly reactive substances so they don't accidentally react with the wrong materials, how the cell "decides" when to end its own life, how it protects itself from attack, and a thousand other wonders. There are even cells that can measure time, and others that can detect electrical and magnetic fields. If that weren't enough, the cell is able to duplicate itself. With all of our modern know-how, what have we made in the physical world that can do that little piece of magic? Yet your body manages to make millions of new cells *every single second*, replacing the millions that die off. Maybe the word "miraculous" is overused, but I'm going to spend one here.

So the cell, with all of this exquisite complexity, seems, well . . . kinda smart. How does it know how to do all that? The simple fact is that it isn't smart, any more than your smartphone is "smart." Just the opposite: Perhaps you have heard the phrase "dumb as a bag of rocks." Well, the cell is just a bag of rocks. I'll say it one final time: The only vital forces in the cell appear to be normal physics and chemistry. Everything we explored earlier—the discriminating cell membrane, the thousands of different proteins that each have a purpose—appears to simply be the most intricate clockwork you can imagine, and yet, as we all know, clocks are not alive.

So is that all life really is? Just a bunch of rocks falling down a hill? Is that all we are? It sure doesn't seem that way, does it? It seems like maybe we are missing something key about the nature of life itself. That's our next stop.

WHAT IS LIFE?

RETURNING TO OUR QUESTION of what exactly life is, let's begin with a dictionary definition, in this case courtesy of *Merriam-Webster*:

> **Life**, noun, **1 a:** the quality that distinguishes a vital and functional being from a dead body **b:** a principle or force that is considered to underlie the distinctive quality of animate beings **c:** an organismic state characterized by capacity for metabolism, growth, reaction to stimuli, and reproduction.

If you notice, though, none of this says what it is. The nouns used are "quality," "principle," "force," and "organismic state." That last one is particularly meaningless. "Rage" and "sleep" are also organismic states, along with a thousand other things. Further, while the attributes of life listed out may be *true* of life, why are they the *definition* of life? Other things can react to stimuli or grow and still not be alive, so why those particular attributes? A definition of "cat" that mentioned it had fur and could purr might describe the cat, but it wouldn't define it. As I pondered this, I noticed there was a medical dictionary on the Merriam-Webster website, so I looked it up there. It's basically identical, with just a couple of words changed. Then I saw they had a legal dictionary but discovered nothing new there. Then I noticed they had a kid's dictionary, and I thought, "That's where the answer will be," but that definition was also virtually identical as well.

Biophysicist Edward Trifonov decided to try to work out a sort of meta definition by aggregating various definitions across not just sources but entire disciplines, and across centuries as well. He collected 123 different definitions of life and attempted to amalgamate them into one. He discovered that "the definitions are more than often in conflict with one another." But he also suggests that "undeniably, however, most of them do have a point, one or another or several, and common sense suggests that, probably, one could arrive to a consensus, if only the authors, some two centuries apart from one another, could be brought together."

His methodology was straightforward. He began by first tallying every one of the words in each of his 123 definitions. Unsurprisingly, he got words such as "complexity," "system," "metabolism," "information," and "growth." Less helpfully, he had frequent uses of the words "live," "alive," and "living." He then took all those words and grouped them by similarity in meaning, giving him a shorter list containing entries such as "matter," "chemical," "evolution," "complexity," and "reproduction," plus a few more. From that, he settled on, "Life is self-reproduction with variations."

While I admire the effort, I still have to ask, "Yeah, but of all the words he had to choose from, why *those*?" You could imagine a life-form that reproduced without variations or that was assembled by something else.

It would be nice if we didn't have to deal with this thorny question in such tedious detail, but our topic insists. We are inching toward questions of whether multicellular creatures are a different kind of life, or not life at all; whether superorganisms are life; and ultimately, whether there are planetwide life-forms such as Agora that may not even have physical bodies. So we must spend some time figuring out just what we mean by "life."

NASA has tried to come up with a definition of life that could be suitably applied to extraterrestrial as well as terrestrial life. They settled on "a self-sustaining chemical system capable of Darwinian evolution." This, too, is an . . . odd . . . definition. Why does it have to be self-sustaining? Chemical? Or, most inexplicably, why must it evolve in such a specific way?

Oftentimes we have difficulty defining really big ideas, such as right and wrong, friends and family, love and hate, home, health, and art. There are pretty good reasons we don't have precise definitions for each of these. Sometimes it is simply because people mean different things when they use these words, or that the sense of the word has much to do with its context. But as I mentioned earlier, this isn't the case for life. We struggle with coming up with a definition for the most obvious reason of all: we don't know what it is.

This is pretty surprising when you consider just how extensively life pervades the planet. It is everywhere, from forty miles up in the atmosphere to a yet-unknown number of miles beneath the earth's surface. University of Georgia microbiologist William Whitman and his team tried to estimate the number of bacteria—just bacteria—on the planet and came up with five followed by thirty zeros. No one can fathom that number, but think of it this way: It works out to about a trillion—an incomprehensible number in its own right—bacteria for every grain of sand on every beach and in every desert on the planet. Or, perhaps more relatable, a single teaspoon of topsoil has more living creatures in it than there are people in the world. Most of this life is utterly unknown to our science. In one experiment, researchers swabbed metal poles on New York subway cars and discovered that half of the DNA they collected wasn't identifiable. With regards to multicellular life—a topic we haven't even come to yet—we don't have any idea how many living creatures there are, or, surprising, even how many different species there are. It is certainly in the millions and may be over a hundred million unique species of multicellular life.

That all seems like it should provide enough examples to come up with some definition. No. Regrettably we only get to count all that as just one single, solitary example since every bit of it, from amoebas to zebras, is descended from what biologists call LUCA, the Last Universal Common Ancestor, a literal, actual, single creature that lived long ago that is the progenitor of us all. Because of it, you have a cousin that is a cactus in Tucson, and another that is an earthworm in Bethesda.

Philosopher of science Carol Cleland, who has worked extensively with NASA and the SETI Institute on questions relating to the nature of life, believes it is a mistake to try to define life, at least yet, stating that "definitions tell us about the meanings of words in our language, as opposed to telling us about the nature of the world." She writes that attempts to define it pose "a dilemma analogous to that faced by those hoping to define 'water' before the existence of molecular theory." Instead of defining life, she believes our efforts should be focused on coming up with a general theory of living systems. However, that is also fraught with difficulty, for to do that we still need more than our lone example of life to work from.

Of course, if we broadened our working definition of life considerably, we may come up with additional examples beyond the one we do have. Biologist Chris Kempes and systems researcher David Krakauer of the Santa Fe Institute have offered an idea of what this might look like, putting forward a "general theory of life that integrates our empirical understanding of biology with logical principles that might transcend it." Their broad definition of life is "the union of two crucial energetic and informatic processes producing an autonomous system that can metabolically extract and encode information from the environment of adaptive/survival value and propagate it forward through time." It's a mouthful to be sure, and it reads like a definition made by a committee, but what it is trying to do is remove the purely biological aspects of life from the list of requirements, which we want to do because it's possible that non-biological things can be alive, too. By this expanded definition, the economy is as alive as *E. coli*, as are cultures and ecosystems.

In the 1994 movie *Reality Bites*, as part of a job interview, Anne Meara asks Winona Ryder to define irony. It doesn't go well. "Irony. It's a noun. It's when something is . . . ironic. It's, uh . . . Well, I can't really define irony . . . but I know it when I see it!" As she relates her bad day to Ethan Hawke, she asks him if he can define irony. "It's when the actual meaning is the complete opposite from the literal meaning." That's the end of the exchange, and it is never pointed out that Hawke's definition was incomplete, simply describing one example of irony—sarcasm. Such is the case with definitions of life. Either we have Ryder-esque ones that boil down to "you know, it's when something is alive," or we have Hawke-esque ones, glib but incomplete.

Technically, there is no "correct answer" to the definition of life, for it means whatever we all agree it means. What we want is a definition that helps us understand just exactly what we are all talking about, one that is narrow enough to be informative but broad enough to include more than just the sole example of our DNA-based version.

Since we don't have an agreed-upon definition, we cheat a bit to get around that annoyance. We don't actually say what it *is*, rather we just make lists of attributes that we associate with living things. It reminds me of the videos of the so-called UAPs, that is, the mysterious "unidentified aerial phenomena" that the US government confirmed have been encountered by the US armed forces. A typical exchange with the media about them goes like this:

> **Reporter**: What are they?
> **Government spokesperson**: They sure are fast.
> **Reporter**: Yeah, I see that. But what are they?
> **Government spokesperson**: And man, they can turn on a dime. And they apparently have no exhaust that we can detect.
> **Reporter**: Yeah, I understand all that. What are they?
> **Government spokesperson**: Did I mention they are fast?

This is sort of where we are with life. The best we can do right now is to cobble together a set of characteristics and traits that we associate with living things. It is altogether unsatisfying because when we are done, we know almost nothing about life itself, rather just a feature list of living creatures. And more precisely, living creatures on Earth. But at least this gives us a sort of checklist for how "lifelike" a thing is.

I've constructed a kind of "middle of the road" list made up of the most commonly cited attributes of life—the ones that appear in biology textbooks, encyclopedias, and dictionaries. Eight characteristics of life appear repeatedly. They are, in no certain order:

Made of cells. While it is straightforward, this one has always struck me as odd since cells are not made of cells. People get around this by playing linguistic games and saying that the single-celled creatures are made of cells, just one. I am unimpressed with this line of thinking, but we will visit it again when we get to multicellular life.

Self-creation. As Fritjof Capra and Pier Luigi Luisi write in *The Systems View of Life*, "Life is a factory that makes itself from within."

Wall or barrier that separates it from the rest of the world. This one is interesting in that it captures the idea that the life-form must be self-contained and clearly demarcated from everything else around it.

Homeostasis. This is a continuation of the last one. Not only is the life-form separated from the world, but it constantly tries to maintain a certain state that is different from it, such as maintaining a certain temperature or pH range.

Consumes energy, expels waste. To do all of the above, the being must have a metabolism, that is, a way to capture and utilize energy. On Earth, all life uses ATP, but how it generates it varies considerably. Some organisms convert light to energy, while others bypass that and just eat other creatures.

Reacts to the environment or other stimuli. This suggests that all life must have senses that allow it to interface with the outside world.

Subject to evolution. The organism has a mechanism to allow for change. This goes hand in hand with reacting to the environment in that evolution provides a learning mechanism for the population's DNA. We will discuss the implications of this in more detail later.

Able to reproduce. This is a tricky one, too. After all, the majority of humans cannot reproduce at all—males, the elderly, children, and so forth—and none can by themselves. Putting that aside, the implication of reproduction is that information can be handed down to subsequent generations. You may light one candle with another, but no information is saved.

While useful as a quick reference, this list has one glaring problem: It doesn't attempt to separate what is *required* for life from common features of living creatures. If you were to list the common features of all cars, your list might include steering wheels, blinkers, and so on. But technically what is required for a car to be a car? That answer may be more related to the essence of car-ness than any list of parts. Maybe we should look at a more philosophical view of life and ask what purpose it serves, what essential function it fills, and what problem it solves.

The physicist Erwin Schrödinger of dead cat fame wrote a book in 1944 called *What Is Life?*, which might seem like an odd question for a physicist to ask, let alone write a book about, except that his answer is based on physics. He focuses on the Second Law of Thermodynamics, the one that says the universe is heading toward entropy, the physicists' word for a lack of order. Schrödinger argues, first, that life's hallmark is that in a universe that is winding down, life is the little pockets of energy scattered about that follow the urgings of the poet Dylan Thomas and do not "go gentle into that good night" but instead "rage, rage against the dying of the light." The second part of Schrödinger's theory was that the genes of living creatures are the mechanism used to resist entropy, seeing them as libraries of successful survival strategies that are improved upon and passed down.

The march toward ever-increasing entropy is one of the unbeatable realities of the physical world. The energy of the universe began orderly and over billions of years is heading to a complete lack of order. There is no way around it, although no one is completely sure why the amount of energy in the universe is constant and why the driving force toward entropy is so relentless. Maybe you and I don't lie awake at night wondering why you can't unscramble an egg, but the big-brain folks scratch their heads over it all the time.

There are two aspects of biology that seem to defy entropy, although neither does. They are life and evolution. After all, you started out as a single cell and now are a highly ordered and complex creature. Where is entropy in that? Likewise, the earliest life was simple and seems to be getting more complex and ordered as well. So what's the catch? The Second Law states that entropy increases over time in a *closed system*. That's the important bit. You, for instance, are not a closed system. You constantly have energy flowing into you through various means, which gives you the power to establish a tiny bit of order and shake your fist at entropy with impunity. But enjoy it while it lasts, for the minute you die, and I mean the very minute, entropy regains its footing and what was recently you quickly becomes part of the dissipating universe.

The other aspect of life that seems to defy entropy is evolution. It, too, relies on a constant stream of energy to increase order and defy entropy. It isn't all that different from our previous example, a single life. An animal passing on its genes to its offspring, which pass them along as well, can actually be thought of as one incredibly long-lived creature regenerating itself on a regular basis. But again, unplug the sun and the party stops real fast.

In Schrödinger's view, life is a logical step in the universe's billions-years-long journey to maximum entropy, each creature a sort of temporary local maximum of a universe running down. This is a harsh doctrine, to say that William Shakespeare and all he wrote was just a little pocket of order that lasted fifty-two years and then disappeared into the beige background of a dying universe. All of the Bard's body heat is still around; it is just spread ever more evenly throughout the cosmos.

But even if Schrödinger is right that a life-form is a pocket of order that resists entropy, we are still missing something. That can't be the whole story because we use energy to create pockets of order all throughout the day, whether by cleaning the kitchen or building a sand castle, but we don't equate those with creating life.

So where does that leave us? Why don't we understand life well enough to define it? I think it is because our manner of inquiry into life isn't suited for the kind of thing that life is. Let me explain.

In the 1940s, John Steinbeck went on a six-week expedition collecting specimens in the Gulf of California with Ed Ricketts, a marine biologist and personal friend. He writes about it in *The Log from the Sea of Cortez*. In that account, he talks about the fish they collected and put in jars of formalin solution. He writes: "It is good to know what you are doing. The man with his pickled fish has set down one truth and has recorded in his experience many lies. The fish is not that color, that texture, that dead, nor does he smell that way."

Steinbeck makes a significant observation in this passage: In the court of science, the only observations that can be admitted as evidence are those that can be measured and quantified. How the fish fought, what it tasted like, and how it smelled are equally true truths as those of its weight or length, but without a way to measure those things, they are ignored. It might be that life's singular characteristics, the sorts of things that would give us real insights into its nature, are not the sorts of things that science counts as evidence.

What sorts of things? One example: Rodney Brooks, the renowned Australian roboticist, observed that if you take a robot, put it in a box, and program it to try to escape, then it will run through whatever programming it has in a calm, sequential order. However, if an animal gets trapped in a box and wants out, it is an entirely different process. The animal wants to escape and frantically tries to figure a way out. That difference is what Brooks terms "the Juice." He thinks we are overlooking something that is fundamental about living systems that makes them different from engineered systems. Brooks doesn't know what it is, but he rejects any suggestion that the Juice is something mysterious outside of normal physics, yet he acknowledges that it could be something that we don't have instrumentation to measure.

Science demands that we ignore our senses, which are our only way to gather primary data about the world, and instead rely on measurements made by standardized instruments. This is not a criticism. First-person experience is both unreliable and unreproducible. If we let that in, well, then, you'd better find a place for Bigfoot on the Tree of Life, because a bunch of people say they once saw one. Yet, in rejecting first-person subjective experience, it does seem like a whole lot of baby is being thrown out with that bathwater. We have decided that science, real science, must be conducted using tools that produce objective, reproducible measurements, and in doing this, much is lost that is undoubtedly true, as Steinbeck noted.

This idea is expanded on in *The Systems View of Life*, cited earlier. The authors say that while this instrumentationalism of science has worked very well for physics, it came at quite a price. This "focus on quantities was extended from the study of matter to all natural and social phenomena . . . By excluding colors, sound, taste, touch, and smell—let alone more complex qualities, such as beauty, health, or ethical sensibility—the emphasis on quantification prevented scientists for several centuries from understanding many essential properties of life." This is a lesson I think we are still in the process of learning.

For a long time, this distinction was purely academic. Before the Scientific Revolution, the world largely lacked the sorts of equipment needed to study even natural phenomena with mathematical precision. The ancients did great with what little they had. Eratosthenes, a Greek librarian living in Alexandria 2,200 years ago, correctly computed the size of the planet using a wooden rod and a piece of string. During the Enlightenment, there came a cornucopia of new devices designed to take ever more precise measurements of natural phenomena. The thinkers of the day went measuring-crazy, and surprisingly, it largely worked, and they unlocked many mysteries of the natural world. Can't this method be applied to understanding life?

Regrettably, no. Science largely follows a method known as reductionism where you try to understand things by breaking them into all their constituent parts and then trying to figure out how everything fits together and interacts. It works really well for simple cases, but it becomes increasingly difficult as complexity rises. You can probably understand how a simple mechanical clock operates by taking it apart, but your smartphone would probably keep most of its secrets intact. Life is staggeringly more complex than a smartphone.

Biology is simply chemistry, which in turn is purely physics, so in theory understanding quantum mechanics should help you diagnose the pain in your big toe. But I say "in theory" because in practice it doesn't. Each higher level of order—from physics to chemistry to biology—exponentially compounds the complexity. And forget about climbing up that ladder even further: Biology should help us understand psychology, which in turn should help us comprehend sociology, which should give us insights into history. All of this is true, but only to a degree, and it certainly doesn't mean that we can understand history by studying atoms, let alone predict the future with them.

Living things are undoubtedly systems, and fiendishly complex ones at that, so it is not surprising that reductionism doesn't unlock the big secrets here. But it's still our main approach, in part because we are so good at it. Plus, because our instrumentation continues to get better each year, we keep holding to the hope that the breakthrough is just one upgrade away. So ironically, inventions like the microscope may have stunted our understanding of life because the temptation is always to just figure out how to zoom in another 10×, and then call that progress. Rinse and repeat. But what do we now understand about the nature of life that we didn't before? Precious little, regrettably.

The biochemist Nick Lane writes in his book *The Vital Question* that biology doesn't have laws as physics does and that "the predictive power of evolutionary biology is embarrassingly bad." He continues, "If we could predict from first principles—from the chemical make-up of the universe—why such traits arose, why life is the way it is, then we would have access to the world of statistical probability again." There had been hope that the genome might provide biology the first principles that Lane refers to, but that didn't pan out, for we learned that the tiny DNA molecule is too complicated for easy disassembly and analysis.

Reductionism does teach us much about biology and life, but the real mysteries, the ones we explored earlier about how living creatures do the wondrous things that they do, can't be understood this way. We probably already know enough physics and chemistry to understand the mechanics of life, but none of that has yet shed light on these weightier matters. There's a Nobel out there waiting for whomever cracks that nut.

But as much of a conundrum as life is, how life came into being is even more of a mystery, but one that demands our time. How do we solve such a riddle, especially one whose answer lies in the distant past, billions of years ago? By peering deep inside our DNA where life's collective history is recorded. That's our next stop.

DNA

BEFORE WE JUMP INTO the origin of life, let's take a chapter and try to unravel some of DNA's mysteries, which will be important to our narrative for two reasons. First, DNA's essential characteristic is that it stores information, and information is one of the pillars that we are going to use to try to sort out how life works. Second, if Agora does exist, then it must have DNA of a sort—some mechanism that allows information to both accumulate and perpetuate.

Quick refresher from biology class: Your DNA contains about three billion base pairs of information, encoded with the four-letter GTCA alphabet we discussed earlier. Your genome is the complete list of those three billion base pairs. Your DNA is divided into twenty-three pairs of chromosomes, or chunks of DNA, and those chromosomes in turn comprise about thirty thousand genes. There's more going on here that needn't distract us: A tenth of your DNA consists of viruses that infected our ancestors long ago; you have cells that are very much part of you yet still maintain their own DNA from eons ago; and so forth. But the big picture isn't all that different than what we learned at school.

But here's where things get interesting. As Matt Ridley writes in his book *Genome,* "Wherever you go in the world, whatever animal, plant, bug or blob you look at, if it is alive, it will use the same dictionary and know the same code. All life is one." That is amazing. What it says is not simply that we are all descended from a common ancestor, but that even today there is just one kind of life, and it speaks one language, the language of our DNA. It is astonishing that the code is so simple, or, perhaps more precisely, so like our own writing systems of letters and words that we can understand it and even describe it in plain English. We had hoped that the instructions encoded in that language would be similarly straightforward, but this was not the case. They are quite complex, and it turns out it isn't just the letters that matter. How the DNA is folded, for instance, also plays a role, as do several other factors.

When you read that we share a high percentage of our DNA with jellyfish, bananas, and mildew, these facts are largely true and can be taken roughly at face value. When you hear that the DNA of any two humans is 99.9 percent the same, this is also true, but it can be misinterpreted, at least by me. I always took it to mean that there's some base-level 99.9 percent genome that we all share that gives each of us ten fingers and toes, and that it has some blanks in it, like Mad Libs, where you can drop in eye color, hair color, and the rest. But this isn't the case. The 0.1 percent where you and I differ is not the same as the 0.1 percent where you differ from any other person. In fact, there is no spot in your genome that you share with every other person on Earth. So there is no such thing as *the* human genome, rather there is Anne's genome and Bob's genome. When you read that 2 percent of our genome is Neanderthal, that should be thought of in the same way, with the implication that there is probably an entire Neanderthal walking around today, but the poor fella's DNA is scattered across a billion people.

If you are curious, identical twins start out having virtually identical genomes. In fact, they can be completely identical, but likely they vary by about half a dozen base pairs, a tiny amount against the three billion total pairs. Occasionally, they may differ by a hundred or more. As they age, the vicissitudes of life alter, or corrupt, their genomes, causing them to further diverge.

Babies have at birth, on average, about seventy genetic mutations, that is, sequences of genes that are not shared by either parent. Most of these are irrelevant, falling in places that either don't matter or matter very little, but every now and then a mutation does something quite novel, for either good or ill. If the mutation is good, that is, offers a survival advantage, then it is passed down to future generations. Bad mutations, correspondingly, are passed down less often. Over the course of eons, advantageous mutations accumulate and give us the amazing diversity of life we see today. Your genome can change by other mechanisms as well. You can get DNA from viruses, or from animals, or even as you read this, your DNA could be mutating because of some stray cosmic particle left over from the Big Bang passing through you.

Human genetic diversity is comparatively low, meaning that the difference between the genomes of any two humans is slight. Chimpanzees, the animal with the genome closest to our own, have substantially more genetic diversity. Two chimps that live on opposite sides of a river are likely more different, genetically speaking, than two people of different ethnicities from different hemispheres. Why are we so alike? Because we have come close to extinction. When the population of a species dips really low, we call that a population bottleneck, and it means that the few survivors have to inbreed in order to survive, resulting in their narrow genome. For instance, everyone living in the Americas in 1491 was descended from about seventy intrepid Paleolithic humans who crossed the land bridge that once briefly connected Asia and the Americas during the last ice age, or so a study done by Rutgers concluded. One controversial candidate for a species-wide bottleneck is the Toba catastrophe of seventy-five thousand years ago, when a supervolcano erupted and brought about a cataclysm that reduced all of humanity to perhaps just a thousand mating pairs. The recency of that event would easily account for our lack of genetic diversity. But we should count our blessings that we are as diverse as we are, for we have nothing on the cheetahs, whose species population dipped so low in the most recent ice age—perhaps just forty survivors—that all cheetahs today are virtually clones of one another. Imagine if we were all clones of one another.

With DNA, little tweaks can have enormous impacts. While we are virtually identical to chimps at a genetic level, we aren't actually all that similar. It isn't, as one might expect, that they look just like us but have green hair. We don't look alike at all. If you ever mistake a chimp for a friend of yours, you might want to consider getting new glasses, or perhaps even a new friend. We really are vastly different. We live twice as long as they do but are physically much weaker. Of course, it is the mental arena where the differences are really highlighted. As Neil deGrasse Tyson put it, "Let's say intelligence is your ability to compose poetry, symphonies, do art, math and science. Chimps can't do any of that, yet we share 99 percent DNA. Everything that we are, that distinguishes us from chimps, emerges from that 1 percent difference." Why is this? How did we get so spectacularly ahead with just a few tweaks to our DNA? The answer may lie in the so-called human accelerated regions of the genome; the forty-nine pieces of DNA of largely unknown purpose that we don't share with chimps.

DNA serves two completely different purposes, either of which seems like it would be a full-time job. First, it is a data-storage mechanism that contains all of the information needed to build and run your body. Second, it is the mechanism of heredity, which is about using that data to create new life. Let's look deeper at both of those, starting with data storage.

How much data is in the human genome? It's pretty easy math. We have about three billion base pairs, so if you wanted to put that in computer terms, it would go like this: Two binary digits—bits—are required to encode each of the four letters in the GTCA alphabet (00 for G, then, respectively, 01, 10, 11), so six billion bits can record your entire genome. Convert that to bytes, then megabytes, and you find that the whole thing, uncompressed, would be seven hundred megabytes. Now that you know all that, you can promptly forget it, because the amount of data in your genome isn't a good correlate to anything at all. A Japanese flower, *Paris japonica,* has fifty times as much as us.

A few years ago, a group of scientists managed to encode a data file— actually a GIF of a galloping horse—inside a DNA molecule, demonstrating its essential function as a data-storage medium. This exercise in miniaturization was not in pursuit of smaller hard drives but something vastly more ambitious. Mastering DNA at the molecular level would have all kinds of benefits. For instance, cells could be programmed to swim through your body logging anomalies that could later be retrieved, analyzed, and corrected.

DNA isn't alive. It's just a bunch of letters, like words in a book. Your body is the computer, the operating system, and the program, all in one. Biology, or at least genetics, is an information science. A cynic might question DNA's motives a bit, noting that it is a molecule whose only ability is to make creatures

that display great drive and ingenuity in their single-minded pursuit to reproduce, for the sole reason of, well, making more copies of the DNA. It isn't
entirely clear whether our DNA exists for our benefit or we for it. One could
argue that that's all life is: an unbroken lineage of data that has been making
ever more copies of itself for billions of years, like a cosmic backup utility inadvertently left running.

DNA's structure lends itself to being accurately copied. Its double-helix
design means that it can be split down the middle, like a zipper unzipping,
and each of the resulting halves will still have a complete copy of the data. The
body can then use each of those two halves as a template to make two identical
wholes, each of which in turn can be split again, ad infinitum. So far, so good.
However, for evolution to work, the error or mutation rate of the DNA copying
has to fit into a pretty narrow Goldilocks zone. If DNA were too resistant to
changes or too good at error correction, then our species would stagnate and
not be able to adapt to changing conditions. On the other hand, if it changed
too much or too easily, then you would get a kind of chaos, and improvements
wouldn't stick.

We don't know what 98 percent of DNA does. It has picked up the name
"junk DNA," although many scientists bristle at the term. Maybe it does nothing at all and is just obsolete genetic data that managed to get itself written
into DNA and then copied ad infinitum. This view is supported by the earlier
observation that the size of the genetic code doesn't seem all that indicative of
anything, unlike, say, brain size.

DNA has many mysteries yet to be unlocked. Somehow very specific information is coded in it in ways we don't understand. Or at least we think it
is coded in there. How do the monarch butterflies know how to make the
four-generation-long trip from Canada to Mexico and then back again, even to
the very milkweed plants that their great-great-grandparents pupated on? Some
lepidopterists suspect that they are somehow performing calculations based
on the angular height of the sun and the magnetic field of the earth, but how
exactly? Crows hate owls so deeply that even crows who are raised in captivity
from birth, having never seen an owl, will completely freak out the first time
they see one. How is that?

It's humbling to consider just how little science knows about our bodies. As recently as 2016, we discovered a new muscle in the human body called the tensor vastus intermedius. A new muscle. That seems the sort of thing we would have locked down a while ago. But that must have been a one-off sort of thing, right? No. In 2020 we discovered a new organ, the tubarial glands. We are still packed full of mysteries. For instance, a Cornell study recently found that children who eat chicken off the bone instead of with utensils are more aggressive and disobedient. A great deal more, it turns out. We aren't sure why. Speculation is that it is something buried deep in the DNA, perhaps related to how the act of baring our teeth was a threat of violence or how when feasting on a prey animal, we had to be on alert to protect our food source from being stolen. But these are guesses.

We have only scratched the surface of DNA. As tempting as it is to spend more time here, we must push forward toward superorganisms and ask our next question: How did life come into being? Understanding the answer to that will hopefully lead us to the origin of multicellular life, then to the origin of superorganisms, and finally give us insights into how a planetary-sized life-form might come about.

THE ORIGIN OF LIFE

LOOKING AROUND AT THE living world, it is hard not to marvel at the variety and, well, the sheer cleverness of it all. Evolution is a mash-up of a curious tinkerer, a crazy aunt, and a mad scientist. It'll try about anything just to see what happens. If you and evolution were throwing back a couple of beers, and you suggested that noses would work better if they were six feet long, then it would make an elephant just to see if you are right.

The litany of abilities that evolution has bestowed on different species is so amazing that they seem like superhero powers. Well, maybe B-list superheroes. For instance, a species of beetle, the fire chaser beetle, likes to lay its eggs in freshly burned wood, so evolution gifted it with an infrared detector that alerts it if there is a forest fire within eighty miles. Eighty miles. On more than one occasion in the 1930s, football games played at UC Berkeley's stadium were beset by them, attracted by the twenty thousand cigarettes per game that were on average smoked by the fans. Next there is the platypus. As if it weren't weird enough already, it also hunts with its eyes closed, detecting prey by the electrical fields in its muscles. We know of a kind of grasshopper that has actual gears for body parts that it uses to lock its back legs together before it jumps. The mantis is the only creature we know of with exactly one ear, not counting my Uncle Glenn who worked in our town's sawmill. And then, of course, there are the octopi. Their wonders border on the unbelievable. Let's spin the wheel and pick just one to marvel at: To adjust to cold water, one species edits its own RNA to be able to function in lower temperatures. That's like you being able to alter your eye color just by thinking about it. We've got creatures with red blood, blue blood, violet blood, green blood, and—just to mix things up a bit—clear blood. In the plant world, there are similar wonders to behold. In Chile there grows a vine that has leaves that perfectly match its host tree, right down to the pattern of the veins, regardless of the kind of tree it is on. This is a life-form with neither eyes nor a brain. And one can only imagine what we haven't yet discovered, for nature has more things we don't know about than ones we do. A group of researchers working in Brazil found a tree stump, a single tree stump like you might have in your own backyard, that had forty-three different species of ants on it, as many as there are in the entirety of the British Isles.

Coupling all of that with the amazing features of both cells and DNA discussed earlier, the question of how it all came to be seems overwhelming. Top of the list of questions, at least for me, are "Why does it have to be that complicated? Why do cells need millions of parts, and why do we in turn need trillions of cells?" In *The Systems View of Life*, the authors, Fritjof Capra and Pier Luigi Luisi, pose these same questions: "The staggering complexity of modern cells, encompassing thousands of genes and thousands of additional macromo- lecular components, elicits the question of whether such a complexity is really necessary for life, or whether cellular life can actually be achieved with a much lower degree of complexity."

Unfortunately, we have many more questions about the origin of life than we have answers. Let's plow forward and look at some of those questions. To orient ourselves, let start with a rough timeline. The earth is about 4.5 bil- lion years old (BYO). Our oldest fossils are 3.5 BYO, but this is a bit mislead- ing since our oldest rocks are only 3.8 BYO. Yet life is probably even older than this. Researchers at Bristol's School of Biological Sciences tried to figure out how old life is by looking at the genome, which contains an unbroken record all the way back to the LUCA, that amazing creature from which we all descend. One of the researchers, Tom Williams, explains that by "combining fossil and genomic information, we can use an approach called the 'molecular clock' which is loosely based on the idea that the number of differences in the genomes of two living species (say a human and a bacterium) are proportional to the time since they shared a common ancestor." Using this analysis, their date for the origin of life bumps right up against when the planet Theia smashed into Earth and formed the moon, an event that would have sterilized the planet, effectively resetting the age of the earth. Yet, surprisingly, just cosmic seconds after that cataclysm, life appeared.

Thus, our first mystery is the obvious one: How did life come into being to begin with? It had to have happened in a process called abiogenesis; which means, paradoxically, life from nonlife. In other words, to state the obvious, the first life could not have been born, rather it must have somehow self-assembled from nonliving, inorganic substances. Based on everything we explored about the cell earlier, with its millions of parts, it sure couldn't have begun that complex, so it had to self-create, then it had to self-improve as well. That's a really tall order, even for a miracle.

That leads to our second big mystery. Since all life on Earth is related, there has been only one time that life formed here, or at least one time that it formed and persisted. This suggests that the mechanism that created life was so extraordinary and uncommon that it has only happened once. That's a bit surprising, since if life formed immediately after the reformation of the earth, you might expect it to be a pretty common occurrence, happening all over the place, at least every few million years or so. But it seems to have happened just once, and at a time when the planet wasn't all that conducive to life. Sure, that's conceivable, but why aren't new types of life-forms emerging all over the planet?

Third mystery: Having formed, why did life remain unchanged for so long? For over three billion years, the planet seems to have been populated just by bacteria and other single-celled life. And those organisms don't seem to have changed all that much in that inconceivably long expanse of time. Contrast that to the Cambrian explosion, which happened about half a billion years ago when every strange Dr. Seussian creature imaginable was prototyped and tested. Why did life stay so static for so long? It wasn't that the earth wasn't changing. We got an oxygen atmosphere during this period, for instance. And the earth virtually froze solid during this time—more than once. Why didn't bacteria form ever more complex and capable organisms, each uniquely suited to its environment?

The flip side of that coin is the problem around multicellular life. While simple cells were spending a whole lot of time doing nothing, what took the more complex cells so long to evolve into multicellular creatures? Either the transition from single-celled life to multicellular life is rare, suggesting that if it took that long for it to happen and stick, then it probably has happened only a few times. Or it is really common, happening pretty easily, in which case why didn't it happen earlier?

Riffing on this is our fourth problem: Why haven't the inner workings of cells evolved? Nick Lane poses this question in his book *The Vital Question*, pointing out that "eukaryotic cells are wonderfully varied in their size and shape, from giant leaf-like algal cells to spindly neurons, to outstretched amoebae." But, he says, everything inside looks alike. "Most of us couldn't distinguish between a plant cell, a kidney cell and a protist from the local pond down the electron microscope: they all look remarkably similar."

Fifth, still along these lines, is DNA. How in the world did that evolve?

Sixth, why hasn't the language of life, the whole GTCA thing, changed any in billions of years? That's a real conundrum. Think about it: Somehow DNA came along very early—we know that because all life uses it—but then it doesn't seem to have ever changed. Sure, it has encoded how to make a wide range of creatures, but the *language* of DNA itself doesn't seem to have evolved at all. This is analogous to humans getting a spoken language, then dispersing around the earth, losing contact with one another. Then three billion years later, someone bumps into someone else from the other side of the planet who is now purple, twelve feet tall, and has four arms; and not only do they understand each other when they talk, they don't even have different accents. So either DNA is prone to changing, in which case we should see more variety, or it resists change, which leaves us having to explain how it got so complicated to begin with.

Seventh is what seems like a mismatch between the mechanism of evolution as we understand it and the complexity that we see. While evolution isn't invoked in how life was created, it is virtually universally held to be the process by which it improves. This belief runs so deep that Darwinian evolution is thought to perhaps be a fundamental force in the universe. You will recall that NASA's definition of life required it.

Today's standard evolutionary theory is basically the same as the one you learned in high school biology: random mutation produces new variations in creatures, then the dispassionate engine of natural selection culls out unfavorable mutations, leaving behind creatures better suited to their environment that then pass these favorable mutations along to their offspring. Rinse and repeat across the eons, and eventually you get the variety we see today.

Few dispute that this happens. The disagreement is whether random mutation is the major mechanism that drives it. That has prompted some biologists to advocate for expanding this narrative to include other major drivers as well, in something called extended evolutionary synthesis (EES). Kevin Laland, professor of behavioral and evolutionary biology at the University of St Andrews, writes in the journal *Nature* about some of the important elements that the standard theory doesn't address: "Missing pieces include how physical development influences the generation of variation (developmental bias); how the environment directly shapes organisms' traits (plasticity); how organisms modify environments (niche construction); and how organisms transmit more than genes across generations (extra-genetic inheritance)."

You would think that suggestions like Laland's would be no big deal, but evidently they are. And while biologists are hesitant to complicate the clean, crisp narrative of evolution with the introduction of potentially dozens or hundreds of other sources for new variations, random mutation may not be enough to account for the variety of life we see around us. In 1980, Douglas Lenat, one of today's big brains in artificial intelligence, published a paper that argued that random selection wouldn't ever yield enough "wins" to make the overall system improve. He drew this conclusion based on efforts to evolve computer programs using similar means. He concluded that while DNA may have begun using only random mutations, it, too, has evolved and now is "an expert program, i.e., one with heuristics for suggesting which (families of) mutations are plausible and implausible." It went from "random generate and test" to something superior: "Evolution itself has evolved by now into a better process, one guided by past experiences, a 'plausible generate and test.'"

Lenat's point is well taken. If natural selection's best trick is still random mutation, then it hasn't learned a lot in the past few billion years. The methods of evolution should be evolving as well, right? If it is still just randomly putting extra toes on things, how long would it take before it just happened to give a Galapagos finch a beak suited to its environment?

There are also timescale mysteries related to evolution. Think about bacteria. They reproduce at varying rates, but we can use three generations a day as a good average. A bacterium born at midnight tonight is a great-grandparent by the end of tomorrow. That's about a thousand generations a year, or a million in a millennium. Meanwhile, Greenland sharks don't reach sexual maturity until age 150. They get seven generations a millennium. Doesn't that give the bacteria an insurmountable advantage that we don't really see in evolutionary history?

Eighth mystery: How did life become intelligent? Ninth: How did life become conscious? Tenth: Why is space so quiet?

And yet, as Jeff Goldblum might say, life found a way. We just have no idea how. We don't know the solutions to these mysteries, and I regrettably don't have much to offer as a denouement. Thomas Edison once said, "We don't know one millionth of one percent of anything," and with regards to life's origins, he might have been generous. The first question alone—How did life form from lifelessness?—sucks all the air out of the room before we even get to the other mysteries. Even that question probably won't ever have an "aha moment" where the answer in retrospect seems screamingly obvious. Just the opposite: the correct answer will need to be so implausible that it will be hard to believe it ever could have possibly happened even once, otherwise you are stuck explaining why it doesn't happen all the time.

Maybe it's not all that shocking that life's origins are so opaque to us. Every example of life we have today, even the simplest, is incredibly complex. Life couldn't really have popped up like that. Imagine if one of the Voyager probes finally made it to a populated alien world, but they couldn't listen to its famous phonograph record because they had standardized on 8-track tapes. What would they deduce about how that craft came to exist? It would be hard to infer the Bronze Age, the invention of writing, the worldwide financial system, the patent system, and the rest of our intellectual evolution as a species with just one piece of advanced technology. But that hasn't stopped us from trying. The candidate theories are numerous, and none of them is the odds-on favorite.

Maybe the most popular is the RNA world hypothesis, which posits that before we had DNA, the world was populated only by creatures with RNA, and biologists have theories on how RNA might have self-formed, stored information, and replicated. Trouble is that we can't make this work in a lab without a great deal of intervention. Others push back on this theory, arguing that it isn't really a scientific theory at all and that DNA may have formed easily from the chemicals at hand. Or perhaps naturally occurring proteins started copying themselves and got all this going.

Then there are what are known as the metabolism-first theories. Fifty years ago, the theoretical biologist Stuart Kauffman suggested a mechanism where life could have naturally arisen based on chemistry. Suppose, the reasoning goes, that some chemical A leads to the creation of B, which leads to C, which in turn continues down the alphabet until somewhere along the way a chemical produces more A, which can be thought of as its offspring, then the process repeats. Kauffman called this circular process an "autocatalytic set." Logically, you can see how this might work. Biologically, it is not so obvious.

About the same time, the Hungarian biologist Tibor Gánti was proposing a similar but more expansive theory. Suppose you had multiple autocatalytic sets interacting with each other. He pointed out that in addition to reproduction, this could store and pass along information. But a final element was still needed: To keep all these chemicals from drifting apart, they are surrounded by a barrier of some kind, that is, a cell membrane. This simple model thus had all the parts needed for the most basic kind of cell: metabolism, genes, and a membrane, all interconnected. He suggested that this is the simplest life-form possible and called it a chemoton.

There are plenty of other candidate theories. The clay hypothesis attempts to bridge the gap between the organic and inorganic worlds by virtue of crystals formed from silicates in clay. It suggests that perhaps substances that naturally grow into crystalline shapes could, over time, be modified by their particular environments until eventually the crystals themselves come alive. This basically leverages the lifelike aspects of crystals—they do grow and reproduce—while minimizing their non-lifelike aspects, namely, that they are rocks. Others credit hydrothermal vents for working some magic—or a particularly fortuitous lightning strike. There is no question that these theories are innovative, but are they true?

I do want to spend time on one less-known theory, because, well, I think it is true. You may be familiar with Moore's Law, the observation made by Gordon Moore back in the 1960s that suggests, in colloquial terms, that the power of computers doubles every two years. Nobody knows why this is, why you can set your watch by this periodic doubling. We have found similar laws relating to the predictable, regular doubling of technology on different timescales, ranging from months to centuries. Two scientists, Alexei Sharov and Richard Gordon, published a paper provocatively titled "Life Before Earth" in which they argue that the complexity of DNA, another technology, also obeys Moore's Law, but it doubles about every 350 million years, not every two. The shocking implication of this is that if you dial this all the way back to the origin of life, you get a date of about ten billion years ago, twice the age of the earth.

Our sun is probably a third-generation star. What that means is that there was a star that ran its entire life and blew up, and that debris became part of a second new star, which in turn ran its course, exploding again, and part of that debris became our sun. How we know this is pretty interesting. Early stars must have contained just hydrogen and helium, because that's all there was in the universe. But over billions of years, fusion inside those stars, at great pressure and temperature, transformed those elements into the heavier ones we have today. Our sun is chock-full of all kinds of heavy stuff that scientists can use to infer its age and generation. It is not as reliable as counting the rings of a tree, but it is pretty suggestive.

What are the implications if life really is older than the earth? Well, it could mean that life began on a planet that orbited one of those earlier stars. When that all ended in a cataclysm, the biological matter on planets orbiting that star must have been blown into what scientists call "smithereens" and largely snuffed out of existence. It is the "largely" part of that sentence that interests us, for it could be that something like bacteria floated through the vastness of space for eons before alighting on our planet and taking root again, continuing its epic story.

This is a pretty old theory, actually. It is called panspermia, and it dates back to at least the nineteenth century and was spoken positively of by Lord Kelvin, who speculated that life came to Earth on a meteorite. So far, however, nothing about this theory is all that provocative. It doesn't answer much, except the questions about how life formed so quickly and was so advanced from the start: It was already old by the time it landed on Earth. But for everything else, it just kicks the can to another planet. You still have to explain how it started there to begin with. Or do you? There's a variant of the theory called directed panspermia. It says that life was *engineered* somewhere and deliberately seeded into space or even planted here. OK, yes, granted, it does invoke aliens in the explanation, but it sure would explain the virtual perfection of the mechanisms of DNA, why it hasn't subsequently evolved, and more.

How could we prove such a thing? In 1973, the chemist Leslie Orgel and the biologist Francis Crick, one of the discoverers of DNA's double helix, published an article offering the suggestion that aliens deliberately seeded our planet with life. In support of this, they offered a couple of observations. First, life emerged on our planet almost instantly, and yet all life on Earth is related, meaning it happened only once, or at least it survived only once. That's ground we've covered. But more interestingly, they reasoned that the basic elements of life should roughly match the basic elements of the environment in which it evolved. Those were, after all, the materials at hand. There is an element, number 42 on the periodic table, called molybdenum, and it is a key part of organic processes on this planet and yet is quite uncommon, being the fifty-fourth most common element on Earth. They suggest that perhaps life began on a planet with a good deal more molybdenum than we have. Biochemist Steven Benner suggests that Mars might fit the bill here, writing that the molybdenum would have needed to be oxidized to facilitate life and that wouldn't "have been available on Earth at the time life first began, because 3 billion years ago, the surface of the Earth had very little oxygen, but Mars did. It's yet another piece of evidence which makes it more likely life came to Earth on a Martian meteorite, rather than starting on this planet."

But what if it came from elsewhere and was deliberate? What could have been an alien civilization's motivation for going to all this effort? Now we are in a realm of pure speculation, but I think the most telling clue would be the answer to the question: "Well, what was the result of it?" Anything that could engineer DNA-based life probably knows what it's doing, so what ultimately happened? Well, at least from our standpoint, the short answer is that we came about, and we have terraformed the planet. Is it now more hospitable to the aliens?

I should point out that Crick went on to distance himself from the theory, later deciding that maybe everything could have been accomplished without appealing to extraterrestrial origins.

Speaking of aliens, we can't really leave the topic of the origin of life without at least speculating on life on other planets. Let's start in our own solar system first. Earth is so drenched in life that it would almost be odd if we didn't find relatives of our DNA-based life on neighboring planets. Earth has regularly been struck by space rocks that hit with so much force that they fling terrestrial matter containing microbial life into space. This life would have no problems surviving the vacuum of space and perhaps reentry to another planet. That's not mere speculation: A lens cap taken off a camera we left on the moon for two years revealed that the bacteria under it were still quite alive, so a trip through space would be no big deal. In fact, the oldest Earth rock ever found was found on the moon. So stuff is always getting kicked around our solar system. There wouldn't be anything particularly profound or worldview shattering in finding evidence of former life in our solar system, assuming it is part of the same Tree of Life that we all sprouted from. In fact, the only interesting question from that would almost be mere trivia: Did it begin there and come here, or vice versa?

What we really care about is the *other* kind of life, unrelated to us at all. What are the odds of that? There are three schools of thought here. One is that the sheer number of planets orbiting the countless stars in the unimaginably large number of galaxies all but guarantees that regardless of how unlikely the formation of life is, however unfathomably improbable, it is still a statistical certainty. Somewhere else there must be life.

The second school of thought also argues for a universe dripping with life, but for entirely different reasons. It argues that life is more or less inevitable given certain environments, such as rocky planets with water. The Nobel laureate and biochemist Christian de Duve wrote a good bit along those lines, concluding that life is a "cosmic imperative." He explained that "life is either a reproducible, almost commonplace manifestation of matter, given certain conditions, or a miracle. Too many steps are involved to allow for something in between." If this is true, then life is abundant throughout the universe, and therefore, this minute there is almost certainly a space battle going on somewhere.

The final school of thought is based on the Rare Earth hypothesis, which posits that our planet is so spectacularly unlikely that no matter how many planets there are, there won't be another one like us ever again. There are all kinds of oddities about Earth, such as how our large moon—an unlikely appendage to our planet—stabilizes our orbit, which moderates our seasons, as well as our plate tectonics; our unusual molten core that gives us our indispensable magnetic shield; and much more. This is all laid out in a compelling fashion in the book *Rare Earth* by Peter D. Ward and Donald Brownlee.

What is the net of this? It's hard: We are dealing with a near-infinite universe and asking about something for which we have a sample size of one: the formation of life. All of our choices seem incredibly unlikely. But, as the English novelist Thomas Hardy wrote, "Though a good deal is too strange to be believed, nothing is too strange to have happened." So we don't know which strange thing actually happened. Arthur C. Clarke once said, "Two possibilities exist: either we are alone in the Universe or we are not. Both are equally terrifying." I don't quite agree; however, if you change that last word to "exciting," then sign me up.

Before we leave the topic of life, we have to look briefly at its evil twin, death. Do things have to die? Let's take a look.

WHY DO THINGS DIE?

WHILE LIFE—CAPITAL *L*—IS BILLIONS of years old, all life—little *l*—is forever young, constantly renewing itself through death and rebirth, cycles that we call generations. In a scene from the movie *Vertigo*, Jimmy Stewart walks with Kim Novak through the giant redwoods of Northern California and says, "Their true name is *Sequoia sempervirens*: always green, ever living." But this is a misnomer, for as long-lived as those majestic trees are, they grow old and die just like you and me.

Why do things die? It seems pretty wasteful. Why go to all the trouble of making a creature just to have it, and all it has learned, die off? If you could make a light bulb that never burned out or a car that never wore out, you would, so why hasn't evolution done likewise? I mean, it has had plenty of time to work on the problem, yet aging is universal, except for a handful of creatures such as lobsters that not only don't show signs of aging, they don't even diminish in vitality with time. Maybe lobsters don't need to age because the natural world does a fine job culling them in the cutthroat world of the ocean floor. Only bad luck can kill a lobster, but all the rest of us have a countdown clock ticking away.

Of course, the obvious reason is that in a world with births and no deaths, population grows unbounded, requiring ever more resources. Not a recipe for sustainability. But that's only half the story. While *we* may regard our own personal aging and natural death as a design flaw, a bug of some sort, it might actually be a feature. For natural selection to work, it needs the old, unselected-for creatures to eventually die off. In other words, the species will only get better once it has killed off the prior generations.

In addition, from a civilization point of view, it isn't just old *genes* that must be swept aside by the verdict of time in the name of progress, but old *ideas* as well. Imagine if the generation that fought the Civil War was not only still alive but, thanks to 160 years of compounded interest, was also fabulously wealthy and powerful.

It is necessary for human parents to live past childbirth age because our offspring are so helpless when young. Grandparents must confer a survival advantage as well, at least at the group level, because biologically we seem to be designed to live that long. But beyond that, evidently there isn't any lasting benefit to great-great-great-great-grandparents, so they are swept aside by the dispassionate forces of natural selection, and thus we die.

Does everything die? Well, I'm afraid that brings us back to the question of "What is life?" What do we say about seeds buried in the tundra by an Arctic squirrel thirty thousand years ago that sprout after being dug up and spritzed with some water? Is that plant thirty thousand years old? Or two weeks old? Were dead seeds brought back to life? Not really, because seeds can actually be dead, unable to perform that trick, so these must have been alive for those thirty thousand years. You see the problem. But there are even more extreme examples of life that have seemingly become frozen in time. Researchers with the Integrated Ocean Drilling Program report finding bacteria, viruses, and fungi well over a mile beneath the ocean floor. They are not dormant; they just consume a minuscule amount of energy. The researchers think they may be *millions* of years old and only reproduce every ten thousand years. Bacteria recovered from water trapped inside salt crystals have been revived after a hundred million years, and scientists are now trying to do the same with creatures from inside of 830,000,000-year-old Australian halite crystals.

Can human life be prolonged? The discouraging fact is that while we get ever better at getting people to 100 or even 110, the number of people who have made it to 125 remains stubbornly at zero. But perhaps it doesn't matter. The real power wouldn't be to live forever, but to be able to choose the moment and manner of your death instead of having it thrust upon you by fate, so that your last words are not those attributed to Queen Elizabeth I: "All my possessions for a moment of time." In the meantime, we have to deal with the uncertainty of our mortality, hopefully seeing it as did the character Ôishi from the movie *47 Ronin*: "None of us knows how long he shall live or when his time will come. But soon all that will be left of our brief lives is the pride our children feel when they speak our names." That's more than enough for me.

MULTICELLULAR LIFE

NOW, LET'S MOVE TO the next level of complexity. We'll leave the topics of life and death in general and look to multicellular life in particular. Our first order of business will be to briefly explore how these relatively new life-forms came into being, then look at how they are a different sort of life than the unicellular variety that was the sole definition of life on Earth for billions of years.

What is the advantage of multicellular life? Specialization and cooperation are often cited as favorable improvements, but multicellularity also brings new problems, such as opening the possibility for shirkers and freeloaders. Cities, we shall see, share these exact same challenges. Hmm.

Conceptually, it is not hard to see how multicellular life might have evolved. A group of identical single cells might happen to cluster together in such a way that it furthers their survival. Over time, some amount of specialization could occur among the individuals, resulting in an organism better able to survive than its single brethren. But it can't be quite that simple. Remember that not only does that have to happen, but also the genomes of the individuals themselves would need to be altered in favor of the new architecture. In other words, ultimately the new organism itself has to reproduce, not simply its individual cells.

As mentioned earlier, billions of years elapsed between the first single-celled life and the first multicellular life. Wouldn't this suggest that it was another near-impossible occurrence? Perhaps, but maybe not quite so much as the formation of life. Biologists believe that multicellularity has evolved and then persisted a couple of dozen times, but most of those were in the simplest organisms. For more complex ones—specifically plants and animals—it has happened only one time each. Everyone agrees the long delay is puzzling, but there are no shortages of explanations.

One suggests that single-cell life took a long time to oxygenate the planet, only then making it habitable for multicellular life like us to come along. Certainly, our oxygen atmosphere changed the trajectory of life on Earth in a way vastly more dramatic than that asteroid that took out the dinosaurs. For a while, we had dragonflies bigger than eagles and millipedes approaching the size of crocodiles.

A second theory says that while it is easy for creatures to come together and cooperate, each time they do it becomes harder for them to later survive on their own, until eventually they cannot. Thus, there is an imperative toward multicellularity that came incrementally but steadily.

A third explanation suggests that multicellularity arose as an answer to predation, which was a biological innovation that came relatively late. Prior to predation, creatures got their energy from the chemicals in their environment or, later, directly from the sun. One day, some creature ate another one and found it to be a dense energy source, and that's when things got real, real tense. In response to this, perhaps the single-celled organisms found that if they were huddled closely enough, they were too big to swallow. Eventually, they merged into a new life-form.

There are many more theories as to the origin of multicellularity and why it was so long delayed, but time doesn't permit us to explore them all. Let's just look at a final one, also related to predation.

Imagine what might happen if a large single-celled creature swallowed a much smaller one, and both survived. Surprisingly, they both might find this an improvement. The little fella no longer has to worry about getting eaten; it is safe within the cell wall of a larger organism. It could then focus entirely on whatever it does best, such as making ATP to power the cell. The swallower might find this to be a better situation as well. Suddenly, it can focus on what it does best and not have to worry about making ATP anymore. This scenario is more than idle speculation; we think this really happened. Your mitochondria, the "power plants of the cells," still have their own DNA, yet they live within your cells. This is not a parasitic relationship but a mutualistic one; the difference is that in the latter, both parties benefit from the arrangement.

That is one way it could work, with everyone keeping their own DNA, the way that newlyweds might hold on to their separate checking accounts. But what about the problem that the organism eventually needs a single genome? Kevin Kelly addresses this in *Out of Control*: "In some cases the genetic strands of two symbiotic partners may fuse. One proposed mechanism for the informational coordination needed for this kind of symbiosis is the known inter-cell gene transfer, which happens at a terrific rate among bacteria in the wild. The know-how of one system can be shuttled back and forth between separate species." This merger of creatures could supercharge evolution. If each one is bringing something different to the table in terms of ability, the resulting organism might do in one fell swoop what would otherwise take millions of years to evolve on its own.

By whatever means multicellularity came about, the resulting creatures were quite different than the cells that comprised them. This is known, as mentioned earlier, as emergence, and it occurs in systems as new levels of complexity are introduced. Emergence will be a key idea going forward, so let's spend a minute digging into it.

EMERGENCE

EMERGENCE IS THE PHENOMENON whereby a system takes on characteristics that none of its parts have. That's another one of those sentences that is easy to gloss over, but it is a big idea, well worth our time here.

Let's start with a simple example. A clock can keep time, but none of its parts can. But there is really nothing mysterious or interesting about that. We understand how it works. But still, timekeeping is an emergent phenomenon of a bunch of gears and springs. Although this is absolutely an example of emergence, we normally think of emergence as something that naturally arises from systems, not something that is deliberately engineered. An example of that would be the flocking behavior of birds. Each bird just does its own thing, and no bird is in charge, but somehow the flock is able to change directions as one and avoid obstacles. Watching videos of flocks of birds weaving and morphing about gives one the feeling that the flock is a thing different from the birds. That's not to say it is alive or intelligent; just that it's different. It has new behaviors. You would think a bunch of birds would just fly around willy-nilly, but there is an elegance to it that we think of as an emergent property. We think we understand how it works, but it is still really hard to model or predict.

We've seen other examples of emergence earlier in our narrative: cells are alive, but none of their parts are, thus life is an emergent property of the cell. Likewise, you are an emergent phenomenon. You have characteristics that none of your cells have, such as a sense of humor. These are examples where we really don't understand how the emergent behavior comes about.

Emergence is often described as "the whole is greater than the sum of the parts," but this is not quite right. Emergence doesn't increase something; it creates something entirely new and often unexpected—or even downright perplexing. Emergence therefore means that the whole is *different* from its parts. That's the point; none of the parts have the characteristics of the new whole.

Emergence is such an old idea that we don't even have to squint to see it in Aristotle, who said, "The totality is not, as it were, a mere heap, but the whole is something besides the parts." Perhaps the word "accumulation" instead of "heap" is easier on our modern ears, but the point is the same either way.

Emergence comes in two forms, weak and strong. Weak emergence is like the flocking behavior of birds. It may be unexpected but not inexplicable. With study, you can understand it, at least in principle. There's no real mystery about it once you figure the whole thing out.

And then there is strong emergence. That's where it is impossible—by definition—to figure out how an emergent phenomenon came about because it cannot be derived from the interactions of the parts. Such phenomena are irreducible, and they cannot be modeled or simulated, as that would require understanding their underlying mechanisms, which by definition we don't. Consciousness is almost always the example of choice to illustrate strong emergence. You could study all the elements in your body until the end of time and never figure out how those materials could have an *experience* of something. Life is another example often used. Strong emergence would maintain that you could spend your entire life analyzing that bag of rocks that is a cell and never figure out what life is or how it comes about. The challenge is that a strong emergent phenomenon is indistinguishable (by us) from a weak one that we have not yet figured out, so we can't know which we are dealing with. It should be noted that many people would claim that strong emergence doesn't exist at all, and it amounts to little more than us explaining what we don't understand using magical thinking. If it does exist—and I suspect that it does—then it is the most mysterious thing in the universe that I know of.

Luckily, we don't have to solve this age-old debate for our purposes. Whether life, consciousness, and superorganisms are simply examples of weak emergence that we have yet to sort out, or whether they are bona fide examples of strong emergence, then the result is still the same: the system's new attributes are mysteries that are deserving of our wonder as much as they are also beyond our understanding, at least for now.

Emergence occurs in both living and nonliving systems. You can see it everywhere once you start looking for it. It's in the schooling of fish, the behavior of mobs, the formation of cities, the political process, and the weather. Traffic jams are emergent phenomena, as are snowflakes, crystals, and the internet. All the social sciences—anthropology, economics, political science, psychology, and sociology—are the study of emergent phenomena related to groups of people. Some particle physicists believe that mass, space, and time are emergent phenomena. I can't even picture what they might have emerged from.

Are there natural laws around how emergence operates? Probably, but we don't know them yet. The so-called laws of nature aren't really laws, for they don't regulate anything. That isn't even a particularly good metaphor for them. They are not prescriptive—that is, they don't tell us what we are allowed and not allowed to do—but are rather entirely descriptive, telling us what regularities have been observed in nature up to now. And the challenge is that we don't understand emergence well enough to have catalogued such regularities and developed them into laws. Lacking laws, the best we can muster is a sort of checklist of properties often cited as being characteristics of emergent phenomena:

Occurs in systems through the interaction of the parts. This is almost definitional. Systems have multiple components that interact. Emergence happens when those interactions produce a hitherto unseen new property. A pile of automotive parts isn't a system, and it can't do anything but sit there. But take those parts and build a car, and you have something with new emergent properties: It can take you from place to place. But it's the same set of parts. What's different? Two things: the pattern, that is, how the parts are arranged, and how those parts interact.

Multilevel. Systems can also be layered within each other. A bee is a system created by the interaction of its cells. A beehive is a system created by the interaction of bees. There is no logical limit as to how nested systems can be, and later we are going to consider whether our entire planet is a massive set of interrelated systems.

At every level of complexity, emergent properties remain. Lower systems become the parts of the larger systems. So a bee is an entire system when looked at in isolation, but that very same bee is a part of the larger system, the hive. We don't know where that ceiling is, though. What attributes would ten thousand hives exhibit?

Bottom up, not top down. Emergence occurs in systems from the bottom up. It is the interaction of the lower-level parts that defines the system's key attributes, and no one is in charge. Think of your body as a system. No one is in charge of it, not even you. You aren't directing the billions of things happening in your body right now. If you cut your finger, "you" aren't directing the platelets to the cut to stop the bleeding. Even your brain isn't doing that, nor your arm, nor the finger, nor even the cut. The platelets are handling it all on their own.

Autonomous agents. In the same vein, emergent systems often have autonomous agents doing the interactions, in both biological and mechanistic systems. You can think of a bee as a little individual, a biological robot of sorts.

Simple programs. As a general rule, what the parts do is usually simple compared with the whole. In fact, simple is often better. The behavior of the system can be fantastically complex and nuanced with just a few different types of agents doing relatively simple things.

Feedback loops. Most emergent systems have a feedback loop that informs the behavior of the agents.

Unable to be anticipated. The interactions of the parts often iterate and compound, making predicting emergent capabilities difficult, but—except

for the case of strong emergence—not technically impossible, just compu-
tationally challenging.

All levels subject to natural selection. Does evolution try to optimize the
individual bee? Or the entire hive? Both. The bees must evolve to survive,
but so must the hive.

We can't leave the topic of emergence without talking about consciousness,
which is regarded by many as an emergent property.

CONSCIOUSNESS

OFTEN IT IS SAID that we don't know what consciousness is, but this isn't quite true. We know exactly what it is: It is the experience of being you. A rock doesn't experience anything, but you do, pretty much all the time. What we don't yet understand, where all the mystery lies, is how simple matter can have *any* experience at all. We don't even know how to scientifically ask the question of how mere matter can have a first-person experience, much less what the answer would look like.

Consciousness is the single thing that makes life worth living. Without it, we would all be unfeeling robots. The thermostat in your home can measure temperature accurately, but it will never have the sensation of warmth. It can't have the experience of being warm, or content, or happy. But this is no mystery. The mystery is how you can feel those things.

The reason the question of consciousness is so irksome is not that we don't know the answer to how it is that we experience the world—science is full of mysteries—but that our scientific understanding of the universe seems to preclude it from even being a rational question at all. It isn't that the phrase "the rock is happy" is simply nonsensical, but that the very idea of happiness seems contrary to a universe made entirely of what the ancient Greek philosopher Democritus termed "atoms and the void."

If the big mystery is how consciousness can exist, then the first possibility we should consider is that perhaps it can't. Maybe you aren't conscious; maybe you only have *memories* of having experiences. In other words, it could be a trick of the brain, and as we have seen elsewhere, that three pounds of goo in your head has an agenda all its own, and it has no qualms about lying to you to achieve its goals. Maybe matter can't experience anything but can have vivid false memories of experiencing something an instant ago. Those two things are different. Although they seem exactly the same in retrospect, they would tell us quite different stories about the nature of reality.

We know that things aren't exactly as they seem. When you stub your toe on the bedstead, you have an unequivocal experience of your toe hurting. And yet, your toe doesn't really hurt. The sensation of pain is 100 percent in your head. But your brain, not wanting the blame for the toe's pain, says, "It's the toe that hurts, not me."

You have probably heard of phantom-limb syndrome, where people who lose an arm or leg sometimes feel pain in the appendage they no longer have. Obviously, their missing arm does not actually hurt, and yet somehow they still experience it hurting. Even people born with missing limbs sometimes feel such pain. In those cases, at least, the pain is "in your head," not in your limb. The brain is giving you the illusion of a pain that you demonstrably cannot really be experiencing in the place where your brain says you hurt. Thus we know for certain that the brain is capable of such a ruse and must ask if this is the exception, or if all our experiences are essentially phantom-limb syndrome.

I don't have a logical argument out of all of this, but that needn't distract us. If we adopt that kind of nihilism, then nothing is knowable. Where does it stop? If your brain is a constant unreliable narrator, then you can never know anything at all to be true, for it might just be a trick of your devious brain. So we have little choice but to give the brain the benefit of the doubt on this one and assume that we are in fact experiencing the universe, not just being duped.

OK, then, if it is such a big mystery why matter can experience the universe, then the most likely explanation is that we just don't understand all that much about the true nature of matter. I find this explanation plausible.

I wrote a book called *The Fourth Age* around the question of whether computers can be conscious. In it, I grouped all the different theories of consciousness I could find and found seven. That's a pretty good hint that we don't really know.

Why does this matter for our purposes? I wish we could ignore the question of consciousness, for it sure complicates things. But we can't. The difference between a life-form without consciousness and one with it couldn't be starker. You can get a sinus infection and take a couple of antibiotics and kill a billion creatures that don't even know they are alive, and no one blinks. But if you kill a billion things who love their lives, then you're a monster.

Consciousness matters because it is inextricably linked to the mind, the self, free will, sapience, and sentience. It's the whole ball of wax. It is what makes life worth living. Take your consciousness away and you are a robot, a zombie, with no experiences of anything.

So we can't dodge the consciousness question, which is unfortunate because, at least at present, we cannot know whether something is conscious. Is a bee? Who knows? Even the bee may not know. What about the beehive? Maybe. Perhaps. There is an old tradition that it is conscious, and that that emergent entity knows exactly who its beekeeper is. Because of this, it was believed that if the beekeeper died, someone needed to go out and tell the hive.

As for creatures similar to humans, maybe we can make an educated guess. If we are conscious, then maybe animals with similar nervous systems are, too. Thus if an animal looks like it feels pain or pleasure, then we should assume it does in fact experience the world. Or we can study animal behavior and make inferences as well. For instance, it has been claimed that elephants have religion and that they worship the moon, which, if true, sure seems like the sort of thing that would imply some level of consciousness. In a 1977 paper called "Religious Behavior in Animals and Man: Drug-Induced Effects," the author, Ronald K. Siegel of the department of psychiatry and biobehavioral sciences at UCLA, quotes from sources in antiquity, including Pliny the Elder and Aelian, that "at the waxing of the moon elephants would gather long branches from the forest trees and in adoration would lift them up in their trunks as homage to the queen of night." We have to suppose there is at least some creative interpretation going on here that conjures up such specific images as "the queen of night," but Siegel also claims that elephants inter herd mates with grave goods, writing, "Other ethologists have observed elephants burying their dead with large quantities of food, fruit, flowers, and other colorful foliage!"

If consciousness arises from the brain—and I stress the word "if"—then it has been suggested that our level of consciousness is determined not only by the number of neurons or synapses in our brains but by their proximity to each other as well, along with the frequency of communication. Intuitively, this makes sense. If every neuron in your brain were scattered around the globe and each were given a pile of stationery and stamps so they could write each other, it seems unlikely that would give rise to you.

I mention this in particular because while a bee has relatively few neurons, they are packed incredibly densely, more so than a bird's. So maybe the bee does have something going on upstairs. There used to be a thought experiment that went like this: Suppose someone were blind their entire life and you gave them three objects of different shapes, say a sphere, a cone, and a cylinder, and taught them the names of those shapes, letting them handle them as much as they wanted. Could they later, upon gaining vision, identify them by sight alone? I say this *used to be* a thought experiment because now we have real examples of that happening, and the answer is no, they cannot. It turns out that whatever mental model of the world that a person blind from birth has lacks the ability to "picture" the shape in question. Of course, to a sighted person, feeling an object while blindfolded, then picking it out of a lineup, is super easy, because we form a mental image of the world. It turns out bees do, too, at least according to research out of Queen Mary University of London. Imagine this setup: Some tiny cubes, like dice, are hollowed out and filled with some tasty bee treat. A different shape—say, a tiny pyramid—is likewise hollowed out but filled with a terrible-tasting substance, like bee cough syrup or something. Turn off the lights, let the bees in, and give them time to walk around in the dark to find the dice and discover that they are full of good stuff, as well as to discover the little pyramids and how awful their contents taste. Gather the bees up, move the shapes around, turn the lights on, and let the bees in again. Do they know to fly straight to the cubes, which they had only felt before but not seen? Yes, it turns out, suggesting they also have a mental model of the world in their heads. That doesn't sound like much, but that's a pretty big deal to brain scientists, some of whom suggest that's a step toward consciousness.

Other people have suggested that the idea of a self implies consciousness; one test for that is called the mirror test and was first offered by Gordon Gallup of University at Albany's psychology department in 1970. It is brilliantly simple: when something is asleep, sneak in and paint a red dot on its forehead. Also, place a mirror nearby. When the animal wakes up and sees itself in the mirror, does it try to wipe the red spot off? If so, Gallup maintained, the animal must have a sense of self, and that it is seeing a reflection of itself.

It's actually a hard test to pass, and very few things can. I know I failed my first three tries. OK, just kidding, but it truly is a conceptually high bar, and the few species that can clear it include dolphins, chimps, elephants, and magpies. Some apes have failed, but this may be because of issues relating to eye contact in those species—they may simply have not looked that "other ape" in the eyes. Cats fail, or perhaps they just don't care. Dogs fail, too, but again, this might be related to the structure of the test. When it is restructured not as a visual test with a mirror but a "sniff test of recognition," the dogs seem to pass. Does the test prove the animal has knowledge of itself? Maybe not. Critics point out that all the test shows is that something can identify its body parts. Everything, they argue, has a sense of self and other. A dog can tell whether it is drinking water or if the dog next to it is drinking water. Thus, if you paint a red spot on a dog's paw and it licks it off, what's it really proving? That the dog has a theory of mind? Or simply that the dog knows the red dot is on its paw?

Finally, others believe in panpsychism, that everything is conscious, at least a bit. By everything, they mean rocks and even atoms. A milder version of this suggests that all living things are conscious.

Because consciousness is about the experience of something, you can only be entirely certain of your own consciousness. We can't test for it. We can reasonably assume it in creatures like us that react to pain the way we do. That may seem obvious to us, but veterinarians trained in the US before 1990 were taught to ignore the cries of pains from animals as it was believed they didn't feel pain. Likewise, until the late twentieth century babies were operated on without anesthesia for the same reason. Yes. You read that correctly. The standard of care until recently—probably within your lifetime—was to operate on baby humans and all animals without anesthesia because they were believed not to feel pain.

Beyond just observing responses, we don't have a way to gauge consciousness, so we don't know which animals have experiences, or whether plants do. This is troublesome for our purposes because we are eventually going to ask if Agora is conscious. How would we know?

Moving past the "How is it that we are conscious?" question, we have to pose a related one. *Why* are we conscious? Why in the world do we need to experience the world? How would we answer such a question? A good starting place is to ask what is different in the world because we are conscious. You could argue that consciousness isn't that useful at all, that a robot version of you with all the same goals and aversions could probably navigate your life as well as you do. But, on the other hand, if the robot stubs its toe and thinks, "Well, that could damage my appendage, I should avoid doing that in the future," then perhaps that isn't nearly as effective as the shooting pain you feel when you stub your toe. Maybe Thomas Hobbes was right, that we are just machines with simple programming: seek pleasure and avoid pain. Does that programming lead to consciousness? Or the other way around?

Others see consciousness as an evolutionary freebie, an epiphenomenon, that is, an unintended consequence of the creation of another mental process. This may well be the case, and if so, then the mystery of consciousness hinges on how we define the nature of self. What exactly are you? It's now time to tackle that question.

WHAT ARE YOU?

THE SIMPLEST MULTICELLULAR LIFE-FORMS have only a few cells—maybe just four, or perhaps even two. It's a bit fuzzy because if all the cells are the same, it is considered a colony, not an organism. But this is a hair we needn't split. The most basic multicell organisms are little more than a handshake agreement among the cells to watch each other's back. As complexity rises and the cells take on increasingly more specialized tasks, what was a loose collaboration becomes a true ensemble, an integrated whole. At that point, the parts lose their ability to live on their own, and they go from being individuals in their own right to a part of a larger whole.

You are a multicellular creature of astounding complexity. Interestingly, we aren't sure just how many cells you are made of, which is surprising as it seems like the sort of thing that would be knowable. Not only do we not know, but our approximations aren't even that close to one another, because the way we estimate is pretty simplistic. It turns out that there are lots of ways to guess. You could start with the average weight of a cell and work up to average weight of a human and come up with one number. Similarly, you can use the average size of a cell as your baseline. But cells vary immensely in size and weight. Human egg cells are large enough to be visible to the naked eye of someone with really good eyesight. Counterpoint to that, the smallest cells are sperm cells, which are just 1/10,000,000th the size of the egg. With these sorts of variations, averages mean very little, as can be seen in the tale of the man who drowned in a river whose average depth was six inches. Because of all of this, the estimates of how many cells make up your body vary by orders of magnitude.

But if that weren't enough, there is another challenge in doing a census of your body: What cells should be counted? Just cells with your DNA? Or all the cells that live inside you? By number—not weight—you are as much non-human as human given that your body is home to about the same number of bacteria and viruses as it has human cells. While they are not *you* genetically, they really probably deserve to be counted because they have formed a core part of you from your birth. Breast milk is full of nutrients that babies can't digest, but it turns out to be the perfect food for these beneficial nonhuman organisms that flourish in a newborn's biome, go off with them to college, and die with them in old age.

All this taken into consideration, let's make our best guess as to a number of cells that make you up. A safe, middle-of-the-road consensus is that there are forty trillion human cells in you and forty trillion nonhuman cells. While that looks like 50/50, that's only by count. By weight you are 98 percent human, since bacteria and viruses are so miniscule compared with your chunky cells. If you want your mind really blown, consider this: Your body is home to thousands of different species of microbes, many of which are unknown to science, that probably have around eight million genes among them, dwarfing your paltry thirty thousand. And whatever creatures have decided to take up residence on your body are really unique to you, but they overlap with those people in close proximity to you; so much so that if researchers swabbed your feet, chest, and eyelids, they could match you to whomever you cohabit with a high degree of confidence.

And yet, even though you consist of trillions of cells from countless species, there is just one you. You are unquestionably a single thing: one solitary being, completely discrete and self-contained. But if you think about it, it is more than a little unclear just exactly what you are. If cells are alive, then we have to ask, what is the basis for saying that you are alive? You aren't a cell. You aren't even remotely like a cell. Cells are bags of rocks, but you are a bag of life, a bag of cells. Just what are you then? Are you your body? Your brain? Your mind? Your soul? Or something else entirely?

Returning to Aristotle, we can confidently say that you are not a "mere heap" of cells but "something besides the parts." But what exactly? That's the kicker. Somewhere along the way, "you" emerge and the heap vanishes. Well . . . that's not exactly right. Nothing at all goes away. What really happens, inexplicably, is that the heap remains but you emerge as well. Let's define "you" loosely as that voice in your head, the thing looking out at the world through your eyes, the entity that makes choices about your life, the captain of the ship of your body. In short, you. So what are you? Where did you come from? You clearly have a "self" that we can confidently assert the four-celled algae does not have. But what is it?

Perhaps you remember an old phrase about "running off to join the French Foreign Legion." It was, and still is, a thing. The idea was that it was a place you could start over. When you enlisted in the Legion, they would accept any name you gave them as yours. (I have mine picked out already, just in case I ever need it.) After a certain amount of service, you were discharged under that name and were a French citizen. No matter what you did before, no matter how badly you screwed up your life, that old you was legally dead, and no charges—criminal or civil—could be brought against that troublemaker. So, the legion was a reset button for life, a true second chance. It was a good deal, but it came at a real price.

Isn't that interesting? If the French government said you were no longer Jacques Sixpack, then you weren't. Your identity was defined by the state. They said *who* you were, but that has no bearing at all on *what* you are. On that question, I think there are eight possible answers, with a certain amount of overlap between them all. There's no "right answer" . . . well, there actually probably is, but I just don't know it. You are, really, an enigma. And yet, an enigma we must try to understand. Why? Because we are superorganisms, and so whatever we turn out to be could well be what Agora is as well.

POSSIBILITY 1—A BODY

The first possibility is that you are your body. That's the most straightforward explanation, and what we usually think of colloquially as our self. If you say, "The bee didn't sting me," you are talking about your body as your self. The challenge with this view is that while you live your whole life thinking of yourself as a sort of unchanging "you," your body is constantly being replaced, with cells being born and dying at the rate of about four million a second. Four million. Every single second of your life. This means your forty trillion cells—the full contents of your body—more or less replace themselves every year. Over a seventy-five-year life, that's about three quadrillion cells coming and going through the turnstiles of being you.

The band Kansas released a song in the '70s that had the refrain "Dust in the wind. All we are is dust in the wind." This is more literally true than they probably intended. Speed up time and you would see clearly that there is no singular "you" in the physical sense, because every day a billion or so of your skin cells blow off of you, becoming actual dust in the wind and, in turn, are replaced with as many new ones. But that is just a drop in the bucket compared with the hundred billion other cells you will lose today and the hundred billion that will be made to replace them.

Thus, everything that you are today, from a bodily standpoint, is only about 1 percent of all that will someday go by your name. While our bodies seem static on a day-to-day basis, they are actually ethereal things; there for only an instant, and then gone. You are a transitory being, a snapshot in time. You cannot cross the same river twice, as the saying goes, because it is not the same river, but more importantly, you are no longer the same person.

Not only is there no part of your body that persists and endures, but none of you even lasts all that long. You may have once heard a certain piece of trivia, that a few of your cells are with you your whole life. For example, it is often said that your brain has the same neurons that it was born with. So, doesn't that mean that the material that makes up those cells is decades old? No. This is often misunderstood. Only the *form* of those cells endures, not their substance. The atoms that make up those cells have been replaced again and again. So yes, in a metaphoric sense, in a Ship of Theseus sense, they are the same, the way a Model T might not have any of the original parts that came with it in 1908 but is still considered to be the same car. But that's really just a technicality. In point of fact, every molecule in every cell, even the longest-lived, is replaced every few years.

Maybe you've even heard of prisoners serving long sentences trying out this logic with parole boards, arguing that they are not the same person who committed the crime all those long many years ago. Judges are largely unsympathetic to this reasoning, perhaps not so much on scientific grounds as the fact that that logic would upend civilization, whose law codes, both civil and criminal, are built around the continuity of your person, which it equates with your body.

However, this sort of thinking still regularly pops up. Maryland recently considered a law that would require the review of sentences of people who had been incarcerated as minors after twenty years of time served. One of the sponsors of the bill listed several justifications for it, including the argument that a thirty-seven-year-old inmate is not the same person they were when they were seventeen, writing, "Quite literally he is a different person because all of the cells in his body when he was 17 have died and been replaced by new cells." The comedian Jim Carrey also said something along the same lines: "After a certain time there's not one cell in your body that is that person anymore so you end up just imitating what you did in the old days and the original inspiration isn't there."

All that aside, from a practical standpoint, with regards to our day-to-day lives, we casually think that we are our bodies, but that is a convenience we don't generally push to its logical limit. If you get a haircut or lose a leg, you are not less "you" in any material sense of the word. However, there may possibly be one exception to this rule: your brain. Is that "you"? If we could remove your brain and place it into another person or a robot, would that thing be you from that point forward? That's our next possibility.

POSSIBILITY 2—A BRAIN

Try to recall something you haven't thought about in a long time. Maybe the name of your first-grade teacher, a favorite piece of clothing you used to wear as a kid, the first concert you ever went to . . . something like that. My guess is that you were able to conjure up those memories in a fraction of a second from a storehouse of literally millions (or billions?) of other memories. How did you do that? There isn't a place in your brain that stores teachers, clothes, or concerts, yet somehow, they are all there.

The answer is that we don't know. We don't know how a thought is encoded, how it is retrieved, and why it is that every time you recall it, it is altered ever so slightly. Our earliest memories are undoubtedly somewhat fictional because we have accessed them many times. Additionally, every story you tell rewrites the memory of it. After the nth retelling of one of your favorite anecdotes, you don't actually remember the place and the events, you only remember remembering them. At that point it is all gone, a fuzzy copy of whatever already was a fuzzy copy.

In spite of all of our medical wonders, the brain still holds on to most of its secrets. The best we can muster in our day and age are machines that measure activity in the brain's different regions. But knowing where something happens, or even seeing it happen on a computer screen, tells you almost nothing about what is actually happening, any more than you can understand how a jet engine works by watching an airplane fly overhead.

Why don't we know more about the brain? For three good reasons: first, living brains are locked away in skulls where it is hard to observe them; second, dead brains tell us virtually nothing about how they operated, unlike, say, a dead heart; and third, they are magnificently complex.

While it would be tempting to assume that that last one, the complexity of the brain, is the important one, it isn't. Frankly, it is hardly worth mentioning. We don't even know how the simplest brains work. Consider the tiny worm *C. elegans*, which is about as long as a hair is wide. Scientists love to study these worms and use them in experiments because they are relatively simple and we know so much about them. For instance, their nervous systems consist of about 7,750 connections between exactly 302 neurons. Three hundred and two. That's it. You might as well name each one of them. And yet we can't figure out how those few neurons come together to give *elegans* its mental abilities.

After all, with just 302 neurons, *elegans* can find food, avoid light, reproduce, and live as full a life as a nematode worm can.

It's not that we haven't been trying to figure out how *elegans* does all that. A group of scientists has been hard at work on this exact problem for over a decade, trying to model those neurons and synapses in such a way as to reproduce the behavior of the worm. It is called the OpenWorm project, and to this day, people involved in it say they may *never* be able to achieve this. We just simply don't understand how the biology leads to the behavior. And if 302 neurons are beyond our understanding, how much more so are your hundred billion?

The polymath Noam Chomsky, who has spent the past six decades on the faculty of MIT, talks about how difficult it is going to be to figure out how our brains work. He minces no words: "Even to understand how the neuron of a giant squid distinguishes a food from the danger, even that's a very difficult problem. To try to capture the nature of say human intelligence or human choice is a colossal problem way beyond the limits of contemporary science." Chomsky may well be right. We know little about the inner workings of a neuron, and it might be that each of them is as complex as a supercomputer, with activity down to the Planck scale, that is, the base resolution of the universe. If this is even partially true, then it makes perfect sense why the nematode's 302 are so puzzling, for if a neuron were simple, like a 0/1 switch, then there is no possible way that 7,750 binary connections could encode all that behavior, let alone could have evolved to do so.

OK. So we don't know how the brain works. At least we know what it does. It stores our memories, performs cognition, regulates our bodies, and a hundred other things. But is your brain "you"?

Before we tackle that weighty issue, let's take a minute just to marvel at the brain. It weighs about three pounds and contains around a hundred billion neurons. Each of those, in turn, is connected to ten thousand other neurons, for a total of about a quadrillion connections. It is that network, that neural net, that we are told is the source of all the wonder. Over a hundred thousand chemical reactions occur in your brain every second, and it is jam-packed with thousands of miles of blood vessels.

Our most powerful supercomputers need something like twenty-five million watts of electricity to run, enough to power a small town. Your brain runs on twenty-five watts, yet it is more powerful by virtually any metric.

As marvelous a thing as your brain is, it's still a hunk of squishy tissue that bears no physical resemblances to you. Is that really what you are, at your core? When you think of just how amazing you are—creative, funny, empathetic, and everything else your mom always told you—it seems incongruous that you spring forth from three pounds of something that is the consistency of Jell-O. Are you tucked away in there somewhere? It would be odd if you are. After all, you are not a thought, nor information, nor computation, nor even a bundle of memories. Those are the things that are the stock and trade of the brain, and yet they seem to have little to do with the nature of your "self."

So, is the brain the thing that makes you "you"? There is actually pretty good evidence against this theory. Consider the cases where a substantial part of a person's brain is removed. One such example is a hemispherectomy, a surgical procedure where half of the upper part of a person's brain is taken out, usually to stop seizures. It is remarkably effective and has been shown to have no real impact on the personality, intelligence, or memory of those receiving it.

But those aren't even the most dramatic examples. Consider those born with brains much smaller than normal because of severe hydrocephalus, a condition involving the buildup of cerebral fluid. Those so afflicted sometimes go through life with dramatically reduced brain matter. For many, it is debilitating, but a substantial number of people who are missing large parts of their brain live perfectly normal lives, sometimes even oblivious to their condition.

Animal studies show interesting findings along the same lines. If you cut off the head of a flatworm, including its brain, it will all grow back, and the flatworm will regain its old memories. How? There's a Nobel with your name on it if you figure that one out. Likewise, the behaviorist Karl Lashley found that if he trained rats to run a maze, then began removing different parts of their brains, he could never eradicate the memory of how to run the maze.

The brain does not seem to be the end-all and be-all of you. I don't want to sound all down on the brain. I have one that I use almost every day, and I swear by it. Literally. But it may not be the entire story. So let's keep looking for the elusive "self."

POSSIBILITY 3—A MIND

Our first two options, the body and the brain, are purely mechanistic explanations for our identity. They both regard human beings as biological machines, solely influenced by plain ole laws of physics. But what if there is more going on than that? What if you are an example of strong emergence, and your capabilities literally cannot be derived merely through physical laws, at least as we understand them? Perhaps, you are really a mind.

What exactly is a mind? Well, the answer to that question is a bit of a dodge, frankly. The mind is everything that you are able to do that seems beyond what any human organ, even the magnificent brain, *should* be able to. Emotion, imagination, judgment, volition, and will are what the mind does, and how do you program that in biological matter? Maybe you can't.

Some people bristle at the idea of the mind. It seems to them to be a pseudoscientific word that is just a stand-in for an even less palatable word: magic. But with all the ground we have covered so far in this narrative, perhaps we can understand the mind as a set of emergent abilities and attributes that come about by the interaction of the cells in your body. If so, then your mind is that voice that is always with you—the thing that you think of as "you."

I deliberately referred to the mind as an emergent property of the *body*, not just the brain. Why would this be true? In what sense are you an emergent property of your big toe, for instance? Try this: Think of a short passage of text that you can recite, like a movie quote or part of a famous speech. Now recite it in your inner voice, not out loud. In what part of your body do you locate that voice? I am guessing it is in your head. Now, mentally try to multiply 25 times 60. Where does it feel that that calculation is happening? Again, you probably say, "In my head." But as hard as it is to believe, that is a cultural feeling, not a biological one. How do we know that? Because for thousands of years, the location of cognition, thought, and memory was hotly debated, and the idea that it occurred in the brain was a minority viewpoint. When the ancient Egyptians mummified someone, they saved all the important organs for the deceased's later use, but they threw the brain away as they thought it was simply there to cool the blood. So when Egyptians thought, it didn't feel like it happened in their head but in their heart.

This is still with us today. The heart wants, we are told, what the heart wants. Memorization is learning by heart. You believe something with all your heart. Generous people are big-hearted, virtuous ones have a heart of gold, timid ones are faint of heart, and sincere ones speak from the heart. Good news lightens the heart; we relay bad news with a heavy heart; exciting news makes our heart leap, while sad news makes our heart sink. Those "I ❤ my schnauzer" bumper stickers would look pretty odd with the heart icon replaced with a brain.

In fact, it wasn't until centuries after the ancient Egyptians that the brain theory was given any credence. It was first offered by Pythagoras, who evidently took a break from triangles to ruminate on the matter. A century later, Plato's assessment concurred with his, but Aristotle's did not. Other traditions put cognition in the liver, which certainly appears to be the central organ of the body. Ancient Judaism says the kidneys are the source of cognition. Still others placed thought in the stomach.

That last idea is actually pretty plausible. You do have a second brain in your gut. Your nervous system is particularly active there. That's why you can "know things in your gut" and why you can get "butterflies" in your stomach when you are nervous. That actually is a real biological sensation, not a cultural one. It's also why mood and appetite are so closely related.

Undoubtedly, we perform cognition and encode memories not just in our brains but throughout our bodies. Our immune systems, as an example, obviously "remember" how to fight off chicken pox. Vision is a good example of something so processor intensive—to borrow some computer lingo—that it makes sense for the eye to do its own thinking on-site. Your retina is actually a chunk of your brain that has grown into your eye, and it transfers data to the brain at about ten million bits per second. Owls are regarded as wise mainly for their large eyes, but they aren't very smart. Their eyes are big because they do so much cognition in them and relatively little in their brains.

Your enteric nervous system, a.k.a. your digestive system or the second brain, has its own set of senses and reflexes and is made up of a hundred million neurons, about the same as a mouse's brain. It also uses the same sort of neurotransmitters as the brain, and in fact there is vastly more serotonin in your digestive system than your brain. Earlier I mentioned that all your cells come together to make just *one* you. But that may not be true. It could be that this second brain is awake and conscious, but it is just not you; it's more like your roommate, one who probably has a running commentary on your eating habits. "Geez," it may say, rolling its metaphoric eyes, "do you really have to put sriracha on everything?" Also, perhaps you sometimes have an intellectual feeling that you should do X, but a "gut feeling" that you should do Y. This is likely your two brains arriving at different conclusions, and "you" have to call the ball one way or the other.

But all of this aside, we usually consider our brain to be uniquely us. Because of this, we think we can hear the voice "in our head." So we usually think of the mind as an emergent property of solely the brain, and that that is our self, while all the rest of our body is meat along for the ride. But you are miraculous throughout, for most of your cognition occurs outside your head. Your body's systems largely act alone, not waiting for a call from "upstairs," which I am certain is how they refer to the brain. Cells live their lives without any instructions from the brain. They communicate with one another directly through the production of proteins.

Recall our friend from a few pages back, *C. elegans*, the tiny nematode worm. Researchers at Princeton University's Murphy Lab recently found a trick that is probably pretty common to other creatures. If you feed a *C. elegans* some bacteria called *P. aeruginosa*, it gobbles it down, regarding it as a particularly tasty treat. Alas for *elegans*, this bacteria makes it sick. As *elegans* lies there suffering, no doubt regretting its life choices, it grabs a bit of the bacteria's RNA, adds it to a certain part of its own hundred-million-base-pair genome, and for the next four generations, its offspring know to avoid those bacteria as well. In addition, *elegans* also seems to have a "jumping gene" that can evidently carry memories to fellow *elegans*, warning them as well. In other creatures, including humans, there is good evidence that genetic switches can be flipped on throughout our lives that specifically allow offspring to inherit the experiences of their parents. This may be part of how phobias develop.

Memory and cognition may be in more places than we have ever guessed. The systemic memory hypothesis posits that all dynamic systems have feedback loops that store information. In other words, they learn. What is intriguing about it for our purposes is that all superorganisms are dynamic systems with feedback loops, as are you. Agora would also be such a system.

One place this can be studied is in organ transplants, because if the hypothesis is true, then memories run all the way down to the cellular level. Do memories of some sort reside inside the cells of the organs that are moved from one person to another? Do people become like the organ donor, inheriting, as it were, some of their personalities or preferences? I know this sounds like a terrible movie plot, where some protagonist trying to solve a crime dies and the person getting his heart is able to pick up the investigation where it left off, or something like that. But that aside, is there any scientific evidence to support the idea that memories are embedded in organs in some way we don't yet understand? The data is inconclusive at this point.

In a paper by Pearsall, Schwartz, and Russek titled "Changes in Heart Transplant Recipients That Parallel the Personalities of Their Donors," researchers interviewed transplant recipients as well as their friends and families, and concluded that there were two to five parallels per case, that is, two to five examples per person where some new behavior was noted in areas such as preferences in food, music, and art, as well as even career choices. They suggest "that cellular memory, possibly systemic memory, is a plausible explanation for these parallels."

Another paper, this one by Mitchell B. Liester of the University of Colorado School of Medicine, also studies the topic. He recounts anecdotal stories, such as a die-hard carnivore getting a vegetarian's heart and subsequently being disgusted by meat, or a transplant recipient developing a new and unexplained interest in classical music and later finding he had received the heart of a violinist. We shouldn't let these stories alone persuade us, of course, and Liester quickly moves past them and spends the remainder of his paper exploring how memories could be stored in this fashion, examining "epigenetic memory, DNA memory, RNA memory, protein memory, intracardiac neurological memory, and energetic memory."

Of course, people become vegetarians all the time, and a heart transplant might be a real wake-up call to cutting out the cheeseburgers. So much subjectivity is involved that it makes disinterested scientific study a challenge. Plus, there are plenty of studies that fail to reproduce these sorts of findings at all. One study out of University Hospital Vienna interviewed forty-seven Austrians who got donor hearts and simply asked them if they thought their personality had changed and, if so, whether it was because of their new heart. Only three said yes to both questions. Yet, three did.

Are "you" your mind? An entity that emerges from the totality of your physical form? Maybe, but we still have five more possibilities.

POSSIBILITY 4—A SOUL

As abstract as the idea of the mind is, it is still an explanation rooted in our physical world, in one way or another. If it is emergent, then it is emerging from our physical form, brought about by the interactions of our many parts. The soul is the next level of abstraction, and if it exists, it has no such tether to our world.

Let's define our terms, even if only briefly and casually. What is the soul? That question has many possible answers, but if we are trying to find common ground, maybe this will do: The soul is the immaterial, noncorporeal part of an entity, usually regarded as immortal. It's what Yoda was referring to when he told Luke that "luminous beings are we, not this crude matter."

How is this different than the mind, which is also immaterial and noncorporeal? Being an emergent phenomenon, the mind—in our understanding—obviously dies with the body. Plus some day we can hope to understand the mind and how it emerges from matter, but with the soul, well, we will probably know as much about it in a thousand years as we do today.

Let me be clear here: We have no evidence for the existence of the soul that can stand up to the scrutiny of the scientific method. Zilch. Zero. None. But as we discussed earlier, that isn't much of an indictment. Saying we cannot *prove* that there is a soul has nothing to do with whether there is one. Likewise, though, the fact that a majority of people on the planet have believed in the soul throughout history is no proof either. History is full of things that "everyone knows" that no one believes anymore.

And yet, attributing the self to the soul, or even claiming that souls exist, is not necessarily an appeal to mysticism, magic, or irrationality. It is not a rejection of science; it is an acknowledgment of the limits of science. Science never claimed to be the arbitrator of all that is true in the world, rather it is something different: a method for refining truths about our natural world. Its methodology is straightforward: Someone has an idea, constructs an experiment, publishes their results, and invites others to reproduce them. If others are able to, the finding is accepted as true—for now—but always subject to reconsideration. Then the next person comes along and moves the ball forward a bit further. Do that long enough, and you get the world of today. The soul isn't a part of that narrative at all, so science is agnostic about it.

No one can convince you that you have a soul or that you don't. So, unlike science, the soul is not a shared truth but a private one. I am certain that you, the reader, already have an opinion on this question, so we can move along to the next possibility.

POSSIBILITY 5—A COMMUNITY

Maybe you aren't a singular "you" at all, but a community of cells. Perhaps you are a group of beings, and that voice in your head is not "you" but snippets of the cacophony emanating from the horde. How is this different from being a body? The body is a unified whole. You are the only one home. The community theory suggests that this is not the case, that you are an amalgam of creatures.

As Kevin Kelly sees it in his book *Out of Control*, "For better or worse, in reality we are not centered in our head. We are not centered in our mind. Even if we were, our mind has no center, no 'I.' Our bodies have no centrality either." In *The Society of Mind*, Marvin Minsky expands on this same idea, writing, "We're always doing several things at once, like planning and walking and talking, and this all seems so natural that we take it for granted. But these processes actually involve more machinery than anyone can understand all at once." He illustrates this by listing out some of what goes into a simple, mundane task like drinking a cup of hot tea. He envisions each of us as a community, full of agents, each with a specialized task. To drink a cup of hot tea, there's an agent that only knows how to grasp things, and it directs your hand to grab the cup. There's another one that specializes in balance and keeps you from spilling the tea, and yet another that's a real pro at moving your arm. They don't have anything to do with one another, and you don't even have to be aware of them; they go about their business and do their jobs. There are thousands, maybe millions, of citizens in the community of you, most of which you couldn't be aware of even if you wished to. There are the ones that run the immune system, others that keep your heart beating, and still others that enable you to recognize the distinctive smell of an overripe banana. And then, when you read, a whole host of these agents is on duty there. One just recognizes the letter A in all its forms. But it has assistants that in turn can only recognize the crossbar of the letter A, and others that can only identify the slanty parts of the letter.

Where are "you" in this setup? Well, you aren't really there. That voice in your head is the chatter of those agents, but you can only make out the individual ones when they shout. If someone bumps you and sloshes the tea, then the agent responsible for not spilling the tea yells, "Ahhh! The tea is about to spill," and that particular agent is, for that instant, you.

Are you a community, not a unity? Perhaps, but maybe you aren't even that. Perhaps you aren't even there at all.

POSSIBILITY 6—AN ILLUSION

In the previous possibility, "you" for the most part vanish as an entity and are replaced by a legion of specialized agents who act independently. This sixth possibility takes this even further. It says you, and that voice in your head, are an illusion—an evolutionary trick created to solve a specific problem.

Here's the reasoning: You have a set of senses that connect you to the outside world. They bring a constant stream of information to your brain. But that information isn't integrated. Your eyes can only see, and your ears can only hear. Imagine for a moment what it would be like if your senses were completely isolated. You would get a visual feed from the eyes, an audio feed from the ears, and a tactile feed from the fingers, but they would be completely unrelated to one another. It's hard to picture, but perhaps you can imagine how disorienting it could be.

The reason we don't experience the world in that fashion is because our brains have learned a useful trick: They take all those feeds and merge them into a single experience of the world, completely integrated. It works so amazingly well that it is almost painful to watch a movie when the audio is just a tiny bit out of sync with the video. So the brain doesn't create a real view of reality but an idealized one—one where the input streams are perfectly in sync, like perfect VR, but, of course, it is really just R.

The side effect of this is that it creates an unintended artifact, the illusion that there is a "you" watching this real-time movie of your activities, about which "you" provide a running commentary. But that commentary is nothing more than the bonus material for the DVD of your life. It's interesting, to be sure, but entirely irrelevant to the story.

"But wait," you might argue. "That proves there is a 'me.' *I* am the person making the movie. And from my director's chair, *I* call the shots." I wish it were that simple; I really do.

In 1999, the psychologists Dan Wegner and Thalia Wheatley published a research paper with a provocative idea: Instead of your making choices and then acting on them, they suggested that the brain works the other way. Your body, running on autopilot, is going through its day, making decisions and performing actions. It doesn't need a "you" bossing it around; its various parts function just fine on their own.

But in this setup, there is this stray mental process, this experience of the integrated sensory stream—what we call the conscious mind—whose whole purpose is to create a narrative, after the fact, of why you did all the things you did. But it doesn't have access to the parts of the brain that made those decisions any more than you have mental access to the part of the brain that makes your heart beat, so it has to make up what seems to it to be the most plausible explanation for the choices that were made. That's why "your" decisions come after your body's actions, not before.

There is a famous set of experiments that show this clearly. In the 1960s, two cognitive neuroscientists, Michael Gazzaniga and Roger Sperry, conducted an experiment on split-brain patients. These were people who underwent an operation to control seizures that involved severing the connection between the halves of the brain so the halves could not communicate with each other. It sounds really dreadful, but it works quite well. Patients who have had this procedure were thus ideal for studying the function of each half of the brain in isolation.

Now imagine you took one of those patients and gave them something like a set of binoculars that could display images: different ones on the right and left side. In this setup, one side of the brain could see one image and the other side of the brain saw a different image.

Speech comes from the left side of the brain, so whatever the subject said out loud only pertained to one of the images. The brain could see both of them, but the right side of the brain can't talk. It does other stuff. So they showed the language side of the brain a chicken's foot and the other side saw a snow scene. The researchers asked them to choose an image for each picture, and showed both sides of the brain a fork, a shovel, a toothbrush, and a chicken. The patient paired the chicken with the chicken foot and the shovel with the snow scene. So far so good. I would have picked those as well.

When asked why they chose those, the speech side of the brain, which saw the chicken foot, said, "The foot is a chicken foot, so it pairs with the chicken." Obviously. They then asked "Why did you choose the shovel?" Remember, the side of the brain that doesn't control speech saw the snow scene. The patient could have said, "To shovel the snow," except that the part of the brain that saw the snow can't talk. The patient could also have said, "I don't know." But what they said was, "You need the shovel to clean out the chicken coop."

You can almost see the non-speech half of the brain thinking, "No, you fool! I picked it to shovel the show!" But it can't talk. So the talking side of the brain is like, "Hmmm. I only saw a chicken, and I picked a shovel . . . but why? Why? Ahh! I know, to shovel out the chicken coop."

This same effect can be demonstrated with other methods, and has been done so countless times. There's even software you can download to easily reproduce it at home with no special equipment.

So this leaves us with the possibility that our brain is mostly running on autopilot, and our conscious mind is just claiming to have made all its decisions, like the annoying manager at work who takes credit for the good ideas of their employees. Only in this instance, the manager is deluded as well, and really believes that they came up with the ideas to begin with.

The strange thing is that evolution has, for some reason, decided to hide all this from us, to make us think that we are calling the shots in our life—that we are making choices. Why would it play such a dirty trick on us? I don't know. Maybe it has to do with issues around agency and individual responsibility. At some level, it would mean none of us are ultimately responsible for our actions because they were done automatically—reflexively even—without any "malice aforethought" and thus weren't actually conscious choices. This may prove to be true, but you can't have a functioning society based on that view.

But if this really is how the brain operates, then the self is an illusion or, at best, a stooge, an imbecile, a patsy, the fall guy who gets blamed for every bad choice the brain makes, even though it had no say in the matter. This is a pretty bleak view and I hope it is not true, but it has to be on our list.

POSSIBILITY 7—A SYSTEM

We've talked about systems a bit already, but we haven't taken the time to define the term. Although it is a word we use casually in conversation, it has a precise scientific meaning. Even more than that, it is an entire academic discipline— you can get a PhD in systems science. In that world, systems are groups of items that interact with each other to produce a unified whole. Examples include cars, pocket watches, and, well, your body. There is no doubt that your physical body is a system and contains many subsystems such as the endocrine system or the circulatory system. The question we want to answer is whether that is the essence of your "self."

The systems model has specifically defined attributes, including input and output, process, feedback, boundaries, and control. Systems can be both living and nonliving. They are agnostic to what powers them or what their components are made of. The core of a system is the arrangement of the parts along with their relationship to one another. Feedback loops try to keep the system within narrow operating parameters. The fact we explored earlier about how the cells that make up our bodies are constantly being replaced is compatible with the systems view of ourselves. Systems do adapt and change, and their parts get replaced. Their continuity over time is their self-sustaining underlying patterns. You may modify your car over the course of a decade, as well as replace many parts, and although it may change considerably over the years, it will never fundamentally change and become a helicopter.

If your "self" really is a system, then it operates in a certain prescribed way, and the essence is the underlying pattern of operation that is preserved over time. It is the thread of continuity that makes you the same person you were a decade ago, even though you are quite different. Somehow that pattern became self-aware, and that is you. That's a thin broth, to be sure, that leaves much unexplained, but at present we are just trying to understand the possibilities, not call the ball one way or the other.

POSSIBILITY 8—A SUPERORGANISM

That brings us to our final possibility, that you are a superorganism. I don't mean it to be the "ta-da!" final possibility that somehow becomes the obvious choice, for it, too, is fraught with questions. I saved it for last because the remainder of this book is about superorganisms. Let's dive in.

SUPERORGANISMS

THE TERM "SUPERORGANISM" IS old, dating from the 1700s, but it has been used in exactly the sense we are using it for over a century. In the 1920s, entomologist William Morton Wheeler wrote a book called *The Social Insects* where he states, "We have seen that the insect colony or society may be regarded as a super-organism and hence as a living whole bent on preserving its moving equilibrium and its integrity." Elsewhere, Wheeler writes specifically about ant colonies, saying, "Like the cell or the person, it behaves as a unitary whole, maintaining its identity in space, resisting dissolution . . . Moreover, every ant-colony has its own peculiar idiosyncrasies of composition and behavior." In that last bit, he is saying that each ant colony has its own distinct personality, which is kind of interesting given that individual ants don't seem to.

Where is the dividing line? When is something a superorganism as opposed to simply a colony or a community of life-forms? At what point does it become a single thing and no longer merely a heap? I have constructed a litmus test for this:

Superorganisms are:

1. Composed of living creatures that
2. have specialized to a such degree that they
3. can no longer survive on their own, and,

4.   through their interactions with each other
5.   produce a new emergent entity that displays novel properties and
6.   that itself has all the traits of a living creature.

As we will see, all six of these apply to both individual humans and bee colonies. Numbers one to five are pretty straightforward, but what of number six? Do both beehives and individual humans satisfy the traits of a living creature? Let's see:

- Made of cells—Yes, if we expand our definition of cells to something like "cell-like organisms."
- Self-creation—Yes. No one builds the superorganism; it creates itself.
- Wall or barrier that separates it from the rest of the world—Yes. It doesn't have to be a physical barrier like a cell wall. It certainly can be, like your skin to you, but it doesn't have to be. It just needs an area that the superorganism regards as "within" and another that is "without."
- Homeostasis—Yes. It has a set of ideal conditions that it tries to maintain.
- Consumes energy, expels waste—Yes.
- Reacts to the environment or other stimuli—Yes.
- Subject to evolution—Yes.
- Able to reproduce—Yes.

What do we say of something that satisfies the first five requirements to be a superorganism but fails on the last one? Where the emergent entity doesn't rise to the full definition of a living creature? That is just a system, not a superorganism. If a beehive fails to satisfy a certain definition of life, then the hive is still an astonishing, emergent thing, but itself is not actually a new life-form.

What about a system that satisfies all of the conditions for a superorganism except the first? In other words, what if it is made of nonliving parts, but the emergent entity does exhibit all of the attributes of life? What is that? Well, definitionally it is not a superorganism, which must be made of other creatures. But is it alive? Absolutely. That is the description of a cell. If it satisfies the conditions of life, then, well, it is alive, regardless of what it is made of. We know of nothing aside from cells that are like that, and we certainly haven't ever engineered anything like it. It's that last condition of life, reproduction, that trips up most human endeavors.

Finally, if something fails both the first and the last, that is, it is not made of creatures and is not alive but satisfies all the other conditions, then it is still a system, and that's what humans are able to engineer quite well. That's a clock.

Some interesting implications flow from this understanding of superorganisms that are worth our consideration. The first is that because superorganisms are emergent entities that come about due to the interactions of their parts, if the parts stop interacting, the superorganism vanishes. It doesn't die, you understand, it evaporates. If all of your cells decided to take a five-minute smoke break, you wouldn't just fall down dead; the emergent being of "you" would cease to exist. Superorganisms don't have any physical reality apart from the action of their parts. Of course, they don't require all the parts to be interacting. Bees sleep just like you and me, but while some of them are asleep, the hive still persists.

That leads into the second implication, which is that superorganisms preserve their identity over time. Individual bees come and go, but the hive maintains a distinct identity throughout. Likewise, as we discussed earlier, you and your body act the same way: Your cells come and go, but you persist. But be careful here. It isn't that you are a different form of life, but that you live on a different time scale. Just like a cell, you, too, grow old and die, just much more slowly. So when you think of how you and your cells share the same matter, try to picture it on two different time horizons. Your cells are rapidly coming and going, living and dying, each surviving just days, while you are slowly doing the same thing over the course of eighty years. I make this point because a superorganism of humans, that is, Agora, would likewise operate on an even longer time scale. From its perspective, we are the cells, living tiny, short lives, coming and going in droves. This means that superorganisms always have dramatically longer lives than their parts.

If the parts of the superorganism used to be independent creatures that have specialized to a degree that they cannot live apart from the superorganism, how did that setup evolve to begin with? We want to understand this to get a sense of how Agora might have come about, and at exactly what point we stopped being a heap of humans and became a superorganism. To answer this question, we need to understand a key concept in evolution called coevolution.

We usually think of natural selection in pretty simplistic terms—the Galapagos finches need specialized beaks to better fit their environment—but it is never quite that sterile, for the process always affects other living things as well. Nothing in nature lives in complete isolation, so each part of the biosphere is always influencing other parts. Sometimes a species partners with a different species and together they coevolve, each influencing the other. They form a partnership, a kind of symbiosis. Do flowers need bees? Or do bees need flowers? Both, of course; flowers get pollinators and bees get food. In a way, you can almost think that flowers are the ones that created the bees, or at least designed them. Flowers came along way before bees, and back in those pre-bee days if a plant wanted to get pollinated, it had to hope for either a stiff breeze or a visit from a stray beetle who just happened to track in pollen on its dirty feet. Bees seem engineered to be the perfect pollinators, right down to their furry little legs that are like pollen magnets. This has to be why nature gave each bee three million hairs, while putting only a hundred thousand on top of each of our heads. So while flowers didn't make the bees, they were there every step of the way, urging them on.

I mean that "urging them on" slightly more than prosaically. The degree to which bees and flowers coevolved is astonishing. Lilach Hadany, an evolutionary researcher at Tel Aviv University, illustrated this using the flowering evening primrose. If you play a high-pitched sound or no sound in the vicinity of the primrose, nothing happens. But play the sound of a bee buzzing nearby, and within three minutes, the sugar concentration in the plant's nectar goes up about 20 percent. Never mind that plants can't hear the way we do; they somehow pull this off, all to make themselves more attractive to the bee and, by extension, to make the bee more successful so future generations can continue to pollinate them.

However, the plant has to be careful. Making nectar consumes scarce resources, and if they make more than they need to get pollinated, much effort goes to waste. Yet, if they are too parsimonious, the bees just start looking in greener pastures for the good stuff, the way my kids would always remember which neighborhoods gave out the best Halloween candy.

But here's the thing: Coevolution can occur within a single species. A bee slightly better at gathering pollen can coevolve with a bee slightly better at tending the brood. They pair up just like the bee and the flower, and over time, each becomes more specialized. That specialization gives them an edge over other bees, and they continue to specialize, but doing so requires tradeoffs: The pollen-pro bee sheds capabilities that allowed it to care for the brood, either because they interfere with its main job or simply are no longer needed. They have no reason not to shed the skills they needed to survive on their own, and so they do. That's how superorganisms evolve from creatures who used to live independently.

But once the superorganism evolves and becomes a bona fide life-form, does evolution shift its focus away from changing the bees themselves to solely work on optimizing the superorganism? This is an interesting question. Life, as we have repeatedly seen, occupies a hierarchy of levels, each more complex than the one below it. No one level can be wholly understood without incorporating aspects of the other levels. As Bert Hölldobler and E. O. Wilson write in their book *The Superorganism*, "Genes prescribe proteins, proteins self-assemble into cells, cells multiply and aggregate to form organs, organs arise as parts of organisms, and organisms gather sequentially into societies, populations, and ecosystems. Natural selection that targets a trait at any of these levels ripples in effect across all the others."

That last sentence is key, for the authors are pointing out that natural selection doesn't occur just at the organism level but at all the levels. Isn't that fascinating? In addition to a superorganism occupying the same physical space as the creatures that comprise it, their evolution is also comingled. It says that not only do bees evolve, but bee colonies evolve as well. The changes that happen to the bees reverberate through all the other levels, including the colony. But likewise, as the superorganism of the colony evolves, it is altering the bees, too. Thus it would stand to reason that a superorganism such as Agora would be evolving along with us, and we in turn are shaping its evolution.

That raises a question: If natural selection is operating on both bees and beehives at the same time, what if those are in conflict with each other? In other words, what if something evolved in bees that was bad for the hive, or vice versa? This actually happens. Bees, for instance, will sting someone who is perceived as a threat to the hive even though it kills them. If we were just talking about the evolution of bees, that would seem like a major glitch. But evolution will favor the level of the hierarchy that is most essential. The hive can live with one less bee, but the bee cannot live if the hive is destroyed. Thus bees act against their own individual interest, but they clearly don't understand this. No bee thinks, "I'll let Jill sting this guy. I'm just gonna lay low awhile."

One cannot help but to wonder if we have interests that are at cross purposes with Agora. If so, what would happen? Well, natural selection would likely favor Agora over us for the same reason as the bees. But given our cognitive abilities, we would certainly notice, right? Normally, sure, so natural selection would somehow have to hide this from us, probably by convincing us that our choices are in our own self-interests. Is it possible for us to be that self-deluded, to be incapable of introspection into our true motives? Yes, it is.

Experimental psychologist Petter Johansson conducted an experiment along these lines. In it, the subjects were shown two photos—basically passport photos—of two somewhat similar-looking people and were asked which one they found more attractive. The photos were then placed facedown, and the one the subject chose was ostensibly slid over for them to pick up and take a closer look. However, using sleight of hand, the experimenters actually slid over the photo the subject *didn't* choose. The subject picked up that wrong photo, and the experimenter asked them to explain why they chose that one. Only about 20 percent of people noticed they were supplied the wrong photo. Only one in five. Of the 80 percent who didn't notice, they constructed new reasons for their choice, which couldn't have been true since they hadn't chosen that photo to begin with. For instance, one person looked at the photo provided to them and said, "I like earrings," as a justification, even though the person in the photo they originally chose wasn't wearing earrings, whereas the one that was slid to them had big hoop earrings. Yet, the time between their selection and then being asked why they made that choice was about seconds.

Johansson believes that much of what we think of as self-knowledge is actually self-interpretation. When the subject lifts the card and is asked why they picked that one, the subject believes—incorrectly—that they chose that one, and there must have been a good reason, so they quickly try to figure out what it was. In the previous chapter, we explored the idea that we make up our reasons for choosing something after we make a choice. That's not new. However, in Johansson's setup, the subjects are found to be doing something quite different: They are rationalizing decisions that they never made in the first place. And they don't even notice this. They actually believe they chose the card they were slid and will reject any idea that they didn't, for when the experimenter would explain to the subject about the sleight of hand and how they were really given the photo they did *not* choose, often the subject simply didn't believe them.

So we may well have behaviors that we are compelled by natural selection to do in service to Agora that we rationalize as being in our own interests.

But even if this is true, these must be edge cases because otherwise the superorganism would never come about. In other words, if the relationship were really one-sided, the parts would never have coalesced into a superorganism to begin with. So coevolution must almost always benefit both parties.

Coevolution, by the way, is unrelated to a similar-sounding term, convergent evolution. That's where evolution arrives at the same solution to a survival problem shared by different species, the way that bats and birds evolved wings independently. This sort of thing goes on all the time. As much as we talk about evolution being creative, it actually recycles many of the same old ideas over and over, but it advertises them as new. The appendix, far from being unnecessary, has actually evolved independently about thirty times, and no creature with that organ is ever known to have evolved it away. It is probably a sort of repository of the gut biome, a backup stash to be brought out if your body ever needs to reboot that system. Surprisingly, nature seems to particularly like crab-shaped animals, evolving them every chance it gets. An article in *Newsweek* titled "Animals Keep Evolving into Crabs, and Scientists Don't Know

Why" reports that . . . well, the headline pretty much says it all. Convergent evolution plays no role in our narrative. I only mention it because it is easily confused with coevolution, which clearly does play a role.

Now that we understand superorganisms a bit, let's see them in action. We're going to spend the next few chapters looking at examples of superorganisms and how they operate. We want to understand how the combination of the simple parts ends up creating a complex organism. We also hope to find commonalities between these very disparate examples in the hopes that they will shed light on whether humans are part of such a creature.

BEES

SOME SPECIES OF HONEYBEES, bumblebees, wasps, ants, and termites are what we term social insects because they live in large, highly integrated communities. These communities attain the status of life-forms in their own right for the reasons we explored earlier, and so we call those life-forms superorganisms. The social insects are a great way to investigate how superorganisms operate because we can observe them more or less at our own physical and time scales, and are even able to directly relate to their various functions, such as gathering food, caring for young, and defending their home.

I grew up on a farm in East Texas, outside of a population-five-hundred town. To be clear, I didn't actually live in that metropolis but was some miles away. I have little nostalgia about farm life, about the chores I had to do in darkness every morning at five o'clock before school. Up until that point, no one had ever even bothered to tell me that there were two five o'clocks in the same day. My parents were loving, wholesome people, but for some reason had a moral blind spot on the issue of child labor, and thus my job was to feed our calves this horrible-smelling reconstituted powdered formula that nearly fifty years later makes me grimace to recall. The only break I got was two weeks in the summer, when I went to Boy Scout camp. There, I would take classes to earn merit badges. One year, I saw that in addition to the usual woodcrafty merit badges that were always on offer, they had a new one, a nerdy one: bookkeeping. Being a nerd, I signed up and eagerly showed up for it the next day. Along with me, there were exactly eight other nerds who had each decided that the most exciting thing they could do at summer camp was to learn office skills.

Then our instructor arrived. He was an old man, so wrinkled there was enough skin on him to make two old men. He was of the no-nonsense, never-smile variety, and the first thing he told us was that there was not, and would never be, such a thing as a bookkeeping merit badge. It was a typo, and we had all actually signed up for beekeeping. One of our nine raised his hand and said he was allergic to bee stings, and the instructor told him that if he got stung, he should just "walk it off." OK, I am kidding with that last bit, but I suspect that is what he would have said.

But serendipity being what it is, I fell in love with beekeeping. When I got home from camp, I ordered beekeeping gear and a starter beehive from the Sears catalog. Where do you suppose I got the several thousand bees to put in the hive? The Sears catalog as well. For $29.99 (this was in 1985). A few days later, my buzzing box of bees arrived.

On hot days, my bees would group together and form a beard outside the hive's entrance, and they would all flap their wings to create a breeze to cool down the hive. That was my first glimpse of the hive as a superorganism, as a larger entity of which the bees were just cells. This particular trick, I later learned, demonstrates a way that the emergent superorganism is actually a different sort of entity from the bees. Bees, for instance, cannot regulate their body temperature, but the hive is able to do so. The bees, it turns out, have mechanisms to heat and cool the hive to keep it at a constant temperature, roughly that of the human body, regardless of the climate where the hive is located.

One of my other chores was to mow our lawn, and the hive was on a sturdy table a foot or so tall in the middle of the side yard. I would mow all around the bees and even under the table, and never once did I get stung. They seemed to not even care I was there. I liked to think they knew me, and while I did steal their honey once a year, I was also the nice guy who brought a bucket of water with sticks floating in it outside the hive during droughts.

Individual bees are not smart. Their brains consist of just a million neurons. You have a hundred thousand neurons for every one a bee has. Its tiny little brain can't do all that much. But the hive is smart, and it can do things way beyond what a bee can do. I want to go into some depth about honeybees because they really capture the mystery of the superorganism. As I describe the seemingly smart behavior of the honeybees, always keep in mind that they don't know what they are doing. They don't have a self with a mind that can reason. Think of them as little mindless biological algorithms.

First, the lay of the land. A hive consists of fifty thousand or so bees. It has one queen whose job it is to lay eggs. She's not in charge of anything. She just lays eggs. Over a thousand a day, which amounts to her entire body weight. The hive also has sterile females that are the workers, and males called drones that serve no purpose except to impregnate the queen. Each year, one or more new queens are born, prompting the old queen to take half the workers and go start a new hive. So not only do bees reproduce, but beehives do as well.

Why doesn't the world get overrun with hives dividing? Because there is a high degree of mortality trying to make a new hive. Before they swarm, the bees starve the queen so she is light enough to fly, then they themselves gorge on enough honey to last them a few days. If it takes longer than that to find a home, the hive dies. And woe to them all if it starts raining. But the old hive can die as well. While it has no predetermined life span, eventually disease enters or it falls victim to predation.

The hive needs a great deal of food, but it has no special knowledge of where the best sources of nectar are. If all the bees just fly out on a whim and guess, it is a poor use of resources. Given that a bee can fly about four miles out—any more than that would require more calories than could be collected on one trip—the bees can cover an area of about fifty square miles, roughly twice the size of Manhattan. How do they cover this territory as efficiently as possible? Ideally, when a bee stumbled upon a good source of nectar, say a cherry tree in full bloom, it would tell the other bees about it, as well as how to get there. The problem is, they don't have language.

Again, what I am about to describe is not the bees thinking, "Man, that's a great tree. You don't see many trees like that anymore. I gotta tell everyone else about this." Bees aren't anywhere near that smart. But they do have their well-known bee dance, which is used exclusively to relay exactly that sort of information. This is usually described as, "the bees are able to tell the other bees where good sources of nectar are." While this is true, it undersells it a lot. First off, how do you suppose the bee indicates the location of the good flowers? After all, the bees are all in an enclosed hive, where it is pitch-black, standing upside down on a piece of vertical honeycomb. How are they going to point a finger and say "over that way"?

What they do is a little dance in a circle over and over again. Picture it around the outside of a clockface. Round and round they go. But occasionally, instead of making another revolution around the circle, they cut across it, bisecting it. Within the hive, there is awareness of only one direction, one frame of reference, and that is down, the direction of gravity. The angle at which the bee crosses the circle relative to "down" tells the other bees the angle relative to the sun to fly to the nectar. In other words, if the bee walks from the six o'clock position to the twelve, that means "fly directly toward the sun." Likewise, walking from nine to three says to "fly with the sun on your left." And so on.

As the bee is walking across the clockface, it does its waggle, in which it wiggles its butt, flexing the muscles it uses to fly. In doing so, it is relaying two additional pieces of information. The number of wiggles is the distance to the good stuff, and the speed of the wiggles is an indication of the amount of nectar there. While walking across our clockface, if the bee slowly does ten wiggles, then that indicates less nectar than if the bee walks to the middle of the clock, gives ten superfast wiggles, and then finishes walking. Same distance, different report.

Interestingly, the only part of the dance that is subjective, that is, the bee's opinion, is the quality of the source. And even that is packed with nuance and subtlety that seems improbably out of place for a creature with a brain smaller than a pinhead. One example: if you go into a field in full bloom and put some dead bees in the flowers there, a forager bee that discovers it will still report back the location, but with less enthusiasm than otherwise, as if to say, "That place gives me the creeps."

The waggle dance is actually a reenactment of the bee's last nectar run. To complicate matters, none of the bees ever see the dance as it is performed in the pitch-dark of the hive. Instead, they feel it happen with their antennae, and, since they are touching the dancer, they are able to pick up exactly what sort of nectar all the fuss is about. So the full message runs something like, "Fly a mile toward the sun and keep your compound eyes peeled for a cherry tree in bloom," except, of course, the bee isn't thinking any of that any more than your kidneys are thinking about how to filter your blood. I know that is hard to wrap your mind around, and we will come back to this repeatedly. But just let that roll around in your head: The bee doesn't know what it is doing.

How do we know the bee doesn't know what it's doing? Easy: A worker bee lives only six weeks. It spends the first two in the nursery, taking care of the brood. The next two have it making honey and performing other activities outside of the core of the hive. Only in the final two does it forage. For the other bees to be teaching the newbie new bee the dance would require a sophisticated language, far more complex than the dance itself. That's quite an investment in a new recruit that has only three hundred or so hours left to live. No . . . a caterpillar doesn't "know" it is making a cocoon to turn into a butterfly, nor does the bee "know" why it is doing the dance; it is just there in its DNA.

For the longest time, it was debated how the bees computed the distance to the nectar, but we have now figured it out. The bee is actually approximating distance by measuring the amount of visual stimuli it receives on the trip. Imagine this setup. You take a paper towel tube, split it open, and paint a bunch of tiny trees all along the inside. Then you tape it back up and put the tube at the door of the hive so that the bees must fly through it to exit. Entomologists discovered that while the bee only flies ten inches through the tube, from its point of view, it just flew through an entire forest. If the scientists put a nectar source right at the end of the tube, the bee would fly ten inches to it, but go back and report that there is nectar a whole forest away. This was confirmed by removing the tube, then putting tiny little RFIDs on the backs of the bees that saw the dance of the tricked bee and went on to overshoot the location by a mile. So the bee is really just a simple little robot driven by cunning algorithms. (By the way, few researchers are more hardcore than bee researchers. Their idea

of a fun Friday night is to chill down an entire beehive to temporarily immo-
bilize its occupants, then carefully glue tiny tracking devices on the backs of
thousands of them, just to get data on which ones do what.)

At any given time in the hive, several bees are doing their dance. All of
them are trying to persuade other bees to do the same nectar run they did. If
they get a few recruits, then those bees make the suggested run, and if it pans
out, they come back and do their waggle, compounding the effect. Then more
bees come back and dance excitedly as well, persuading more converts. Eventually the source is depleted, and subsequent bees come back less enthusiastic,
ending the hoopla about that particular cherry tree.

Earlier I mentioned the annual swarm. In it, the queen leaves with a ret-
inue of several thousand workers to find a new home. They fly in unison to a
temporary home, a base station from which to conduct their search. They hang
together in a ball of bees about the size of a pumpkin on a tree branch or under
an eave. Once settled, about 3 percent of them go looking for their new home.

The number of factors that make a home good or bad is pretty long, and
somehow the bees, even with their minuscule intellects, are able to evaluate
and weigh them all appropriately. The ideal home must have a certain volume,
provide shelter, and have no ants. It needs to be a suitable distance away from
other hives, oriented in a preferred direction, be a certain height off the ground,
have an entrance within a range of sizes, and so forth. When a bee comes back
and thinks it has found a good home, it does the waggle dance, trying to persuade other bees to go check it out. But this is a life-or-death situation. If, while
gathering nectar, a bee sends other bees to a mediocre nectar spot, no harm
done. But pick the wrong home and that's the end of the colony. To account for
this, the hive has developed a robust error-correcting algorithm: Say that a bee
thinks it has found the perfect home, but candidly, it has vastly overestimated
the virtues of the locations it's advocating. It dances its little heart out and
persuades two others to go check it out. They come back, all like "meh," and so
only half-heartedly do the dance. No other bees would be convinced by such a
lackluster endorsement, so the bad idea dies on the vine.

But let's say the bee did find the perfect new home. Then the other bees sent
to investigate will come back and start dancing with passion and conviction.

Still more bees see this and go check it out. Eventually, when there is a quorum of about thirty bees of which about three-quarters are waggling the same location, the swarm is convinced and the decision is made that that will be their new home. It is a sort of bottom-up, error-correcting, problem-solving capability. How does all this come about? We don't know. At all. We can describe the action, but we don't yet understand how the bees perceive all of this.

As cool as that is, the magic is only just beginning. How does the quorum of bees tell the other several thousand they have reached a decision? What is the bee equivalent of the white smoke that tells the world a new pope has been chosen? Bees have this trick, used in the hive to help regulate its temperature, where they can flex their muscles a certain way to raise their body temperature. They do this—as if in celebration—when the hive settles on a new home. As the other bees feel the bees around them heating up, they get caught up in the excitement and heat up as well, until the swarm itself hits some temperature threshold, at which point it takes collective flight. The bees that know the new location repeatedly fly through the cloud of bees in a straight line pointing the right direction. The other bees start going that way. Meanwhile, bees that know about the new location are already hanging out there, laying down scent indicators to assert their ownership and flag down their comrades. How does all this happen? We don't know.

I go into all of this detail to make a simple point. No bee in that swarm has ever swarmed before. The swarm is annual, while every bee, except the queen, lives only weeks. All that complex swarm behavior therefore could not have been learned, meaning it must be coded in the bees somewhere, likely in their DNA.

Bees aren't the only superorganisms in town. Let's take a closer look at some other examples and see how they operate, looking for common themes.

ANTS, TERMITES, AND THE HUMAN BODY

I KNOW FAR LESS about ants than I do about bees. Regrettably, there was no merit badge class on ant keeping that was mistyped as "accounting." But I do know that they also exhibit amazing behaviors as a superorganism.

Ants are doing pretty well in the world. Collectively, they outweigh humanity, meaning there are millions for every one of us. They are much like bees in many ways. They have a queen that only lays eggs, sterile female workers, and males that exist only to impregnate the queen. In some species, the males don't even have jaws with which to eat, a not-so-subtle hint that the colony doesn't plan to keep them around long enough to work up an appetite. Ants propagate their colonies by sending out a new queen with some workers. The queens live for years, even decades, while the workers have short lives. And, like bees, somehow the overall colony has a long-term memory that exceeds the lives of the individuals.

Among the thousands of ant species, leafcutters are a definite fan favorite. They are amazing. They live in giant colonies that are up to twenty-five feet deep, and each colony can have millions of ants and persist for decades. They are one of a few hundred species of ant that survive by farming fungi to eat. The leaves that they are named for cutting are food for the fungi, not for the ants. The fungus that they farm—and yes, "farm" is the most accurate term for it—is now so specialized that it couldn't grow on its own, much like some of our highly domesticated crops. In fact, when a young queen goes off to start a colony of her own, she takes a piece of the fungus with her, sort of like a sourdough bread starter, to plant and raise in her colony. The fungus is completely cut off from other sources of genes, so over time it becomes ever more dependent on the leafcutters, as they in turn become more dependent on a fungus.

Here's what's even more interesting. If it starts to rain hard, a few leafcutters head over to the mounds' entrances to plug them up with leaves and clay, lest the nest flood. The problem is that the nest produces a large amount of carbon dioxide, and those levels then rise in the now-plugged-up nest. This is no problem to the ants—they have no trouble with that atmosphere—but too much CO_2 will kill all the fungi. So as the CO_2 levels rise, the ants have to open the doors to let fresh air in, but then they have to quickly close them to keep water out. Think about that: Leafcutters have never heard of CO_2, nor do they have any knowledge of the respiratory needs of fungi. How do they know to do this? Perhaps natural selection killed off all the colonies that didn't know this trick, and now we are left with a bunch of leafcutters who are like, "Well, I guess I'd better open the door and let some fresh air in, but for the life of me I have no idea why."

Termites in Africa build similarly impressive homes. Their nests feature ventilation shafts that keep fresh air moving through the mound, and they handle the flooding problem by building giant reservoirs in their nests through which they route rainwater using gravity. Again, no termite knows what convection is, or even air for that matter.

What's even more amazing is that in both cases, there isn't some smart ant or smart termite who is in charge and directs the building to a certain plan. No one tells a worker ant what to do. They are autonomous bots. If they encounter a job that needs doing, like a pebble that needs moving, they just do it. The ants who, in the rainstorm, plug up the holes in the mound aren't specialized to that task, rather they just happened to be near the entrances when some part of their programming was activated by a certain amount of moisture, e.g., rainfall. Because it looks so organized and efficient, we naturally assume there is an intelligence governing it all. Even though I know that they are just tiny automatons, I can't help but project onto them some will and some underlying plan.

Even their building projects are similarly ad hoc. How do a bunch of ants acting on their own produce such exquisitely engineered homes? That's the mystery, and whatever answer there is to that question is also how the emergent colony comes to be.

We have some ideas of the mechanisms that make ant colonies so smart. For instance, every time two ants cross paths, they give each other a quick sniff, which tells them quite a lot, such as what job each of them is doing. The ants are "programmed" so that if they meet too few doing any particular job—say, foraging—they start doing that job. It is a simple rule, but it means that the distribution of jobs is always error correcting for the needs of the group. It is also an algorithm highly resilient against bad data. Say there is a shortage of foragers, but one ant happens to run into several of them in rapid succession— giving the appearance that there are plenty. That ant will act on that data incorrectly, but because there are still many other ants that didn't get the incorrect information, it doesn't matter much.

There is a fascinating parallel with this and human society. During and after each world war, the number of male births spiked in all the participating countries, both winners and losers, for reasons no one knows. It is called the returning soldier effect. It might be that our bodies have similar capabilities that we are unaware of, whereby every time two people pass each other, they make a mental note of the gender. If there are too few men relative to women, does our biology pick up on that somehow? Perhaps, but many other theories are offered to explain this puzzling fact as well.

Another simple but effective rule works like this. Each queen termite constantly produces a pheromone that blocks the reproductive development in the other termites, so there is only one queen. But, when the queen dies, the pheromone stops being produced, and—voila!—a new queen can emerge, which then starts producing the pheromone. Likewise, if an ant finds a good-sized piece of food, like half a hot dog, it lays down a whole lot of scent there, which attracts a great deal of attention. But as the hot dog is carried away, bit by bit, it's no longer such an impressive find, so fewer ants lay down scent, and what was already there naturally dissipates, allowing everyone's attention to turn elsewhere.

Likewise, when ants die, they give off chemicals that signal to the other ants to carry the corpse away to the ant graveyard. There really is such a thing. One scientist took some dead ants, isolated that chemical, and sprayed it on a living ant. Whenever it would try to get to work, a few other ants would haul it off the graveyard, even with it kicking and screaming, "I'm alive, you fools." Every time it would leave the graveyard to get back to work, a new batch of ants would haul it off again. That shows that these are simple machines, programmed to do a narrow range of activities based on a narrow range of possible inputs. It also shows how fun science can be sometimes.

In old Westerns, there was often a hero who got shot and, as he lay there injured, would tell his friends, "Go on without me." Of course, in a white-hat Western, the friends would never leave a comrade behind, so they would risk personal harm to carry the reluctant patient to medical care. Likewise, if an ant is injured, it, too, refuses help, but with a sincerity not present in our cowboy. If the other ants try to help, the hurt ant fights them off, absolutely refusing assistance of any kind. This behavior is beneficial to the colony: If the ant can't be saved, then any resources spent trying to do so are wasted, and thus bad for the whole. Before we sing accolades for the ant, perhaps you have noticed the pattern here. The ant doesn't know what it is doing. It can't be altruistic any more than it can be selfish. It's just a program. It doesn't care about itself; it doesn't care about the hive. Your cells, by the way, behave likewise. When their time has come, they self-destruct. The few that buck this trend, that defy their fate and live beyond their appointed days and refuse to go peaceably, are called cancer cells.

Do we see other examples of superorganisms? Mole rats seem to be one, or at least one that is in the process of emerging. In a colony of mole rats, comprised of perhaps one hundred individuals, there are specialized roles, like a single reproducing queen, many sterile female workers, only a few reproducing males, and, interestingly, something called a disperser. Over multiple generations, the mole rat colony becomes ever more inbred, which over time would reduce its fitness. So imagine there is a male mole rat, dripping with reproductive hormones, that has no interest in mating with the local queen and instead leaves the colony to find a new one, spreading genetic material outside of the inbred colony.

Finally, moving past the social insects, let's return to individual humans. As a superorganism, your body is not all that different from a beehive. As Matt Ridley writes in his book *Genome*, "The relationship between body cells is indeed very much like that between bees in a hive. The ancestors of your cells were once individual entities and their evolutionary 'decision' to co-operate, some 600 million years ago, is almost exactly equivalent to the same decision, taken perhaps fifty million years ago by the social insects, to co-operate on the level of the body."

Interestingly, when people make lists of superorganisms, humans are often left off. There's no real debate about whether we are or are not superorganisms—we are clearly creatures made up of other creatures that display emergent capabilities. So why are we omitted? I don't know, but I suspect it is because we don't really feel like superorganisms—since the lives of our cells are so different from our own, we don't really draw any sort of equivalence between the two. We are just "us" in our minds, each a single being. The fact that we are made entirely of other life-forms is almost just trivia to us more than part of our daily experience.

A superorganism of all humans, that is, Agora, would be vastly larger than anything we have discussed so far. Is there an inherent upper limit to the size of a superorganism?

PLANETARY SUPERORGANISMS

TO UNDERSTAND A BEEHIVE as a superorganism is pretty straightforward, and even thinking of the human body as one is not much of a stretch. Also, the idea that different levels of life coexist within the same matter, superimposed upon each other—that is, in nested layers, in hierarchies—isn't too much of a leap either. But how many nested layers are we talking about? How big can a superorganism be? Could there, in theory, be a planetary-sized one? Absolutely. With superorganisms, bigger is usually better. Beehives produce more honey per bee the larger they get, and large ant colonies are more efficient than smaller ones as well.

Science has been toying with the possibility of planetary superorganisms for well over a century. The first hint of something like it would be when we got the concept of the biosphere. That term came about in the late 1800s from Austrian geologist Eduard Suess's book *Die Entstehung der Alpen*. The biosphere is composed of all of the parts of the planet where life exists, and he maintained that we could gain great insights into the nature of living things by embracing holism, the opposite of reductionism. Just the notion that there was a biosphere that was a useful scientific concept was considered a bit "out there" at the time.

In the 1920s, that idea was expanded on by Russian/Ukrainian mineralogist and geochemist Vladimir Vernadsky, who saw the earth as consisting of first the geosphere—all the solid parts of the planet; the rocks and such—and then, on top of that, the biosphere: the realm of all living things. He was part of a small group of thinkers who conceived of an additional layer atop all of that, a layer of mind and thought enveloping the earth, which he called the noosphere (NO-iss-fear). He believed that cognition, especially in its amped-up human form, alters both the geosphere and the biosphere.

The core idea was adopted by a number of people who have somewhat different takes on it. One of these was a colleague of Vernadsky's, Pierre Teilhard de Chardin, a Frenchman who evidently had difficulty deciding on a college major and ended up with the improbable résumé of being a biogeochemist, paleontologist, philosopher, and Jesuit priest.

While many of its adherents believe it to be a bona fide superorganism, the noosphere has always struck me as far more mystical than scientific. One website explaining the concept, humanenergy.io, suggests that perhaps "the Noosphere may thus offer meaning and purpose, by guiding and inspiring us toward an amazing planetary transition" and poses questions such as how "could we retain our individual freedom with the coming of a superorganism?" The theory deserves mention, though, because it really is a breakthrough in thinking of a life-form on a planetary scale. But the noosphere is nothing like Agora, which is more like a giant beehive than a great awakening.

This leads us to the Gaia (GUY-ya) hypothesis, which postulates a true planetary superorganism—one also that claims to be entirely based on science. We're going to spend time here because it offers some great insights on how superorganisms do what they do. It was put forth by the English scientist James Lovelock around 1970 and is named for a Greek primordial deity who was regarded as the goddess of the earth. The name was suggested by William Golding, known for writing *Lord of the Flies*, who used to take afternoon walks with Lovelock around the small village where they both happened to live. The simple form of the Gaia hypothesis goes like this: The earth systems, both geological and biological, function as if they were a single self-regulating organism for the purpose of maintaining conditions conducive to life. It sees the earth and all of its life as a single system.

Recall our bee and flower example when we discussed coevolution. They coevolved symbiotically to further their own survival, and in doing so, each furthered the survival of the other as well. Forty million years ago or so, bees and flowers decided to throw in their lot together, and now they prosper or fail together. Kill the flowers and the bees die; kill the bees and the flowers die. They have become a single ecological system.

The Gaia hypothesis imagines that same dynamic on a planetary scale where the earth and all of its systems are the flower, and every species of life are the bees. Wipe out the ecosystem, damage it too much, and all the life dies. But this works both ways: Remove all the life from the earth, and it, too, changes. The surface temperature skyrockets, the rain stops falling, and Earth becomes a miserable wasteland, just another Mars or Venus, just another dead rock in space. The idea is not that the earth just happens to be such an Edenic paradise ideal for life, but rather that life transformed it into one through their symbiosis.

How would you prove such a hypothesis? The most suggestive evidence is the way that certain aspects of the natural world seem to be held within the tight ranges that life prefers, even though by all logic they should fluctuate wildly. Recall how our bees regulated the temperature of the hive: If a few bees get hot, they start working to cool things down and are joined by other bees until the hive is brought back within an ideal range. So whether the temperature outside is thirty or a hundred, the self-regulating mechanism can hold the hive's "body temperature" constant. Life actively alters its environment for the very purpose of making it ideal for life. Lovelock believed that all life on Earth likewise self-regulates the various conditions on the planet, making this a paradise of a planet for our kind of life.

What are these mechanisms allegedly used by Gaia to self-regulate? Let's look at a few of them because they demonstrate some of the techniques by which superorganisms self-regulate.

Surface temperature. Since life began on Earth a few billion years ago, the energy from the sun hitting our planet has gone up by almost 30 percent. That sounds like it should have been the death knell for life. However, throughout that time, the temperature of the earth has continued to fluctuate within a narrow range, suitable for life. We know this for a number of reasons, not the least of which is that life has survived without interruption. Even when the climate has gotten precipitously hot or cold, the regulating mechanisms of the planet have kicked in and restored it to roughly what we have today.

It is thought that many processes are involved in this temperature regulation. Maybe hundreds. One single example: When the planet gets too hot, the glaciers melt. They are so thick and heavy that when they retreat, the ground actually rises. This causes a great deal of volcanic activity that spews ash into the sky, blocking out the sun, cooling the earth. This causes the glaciers to build up, which patiently wait until they are needed again. Other temperature-regulating mechanisms include photosynthesis, the jet streams, the oceans, and even the color of vegetation, which gets darker when the planet needs to warm and lighter when it needs to cool. Of course, all of these mechanisms work on a planetary time scale, not a human one. And it should be noted that Gaia makes conditions conducive to life in general, not human life in particular.

Ocean salinity. The oceans have held their salt levels at more or less 3.5 percent for a long time. By "a long time" I mean something like 1.5 billion years. And we should all be thankful, because if the concentration rose above 5 percent, most cells couldn't survive in it. On top of that, most life couldn't survive if the salinity changed frequently. But how is it held constant? After all, rivers are always washing salt into the oceans, and then as water evaporates from them, they should be getting saltier. There is still debate on this question, but part of the answer is a certain bacteria that removes the salt from the ocean water and sequesters it, creating the vast salt plains we find scattered about the planet. More salt brings about more bacteria, which lessen the salt levels, which causes the bacteria to thin out, until more salt comes into the ocean.

Oxygen levels. The earth's oxygen levels have been holding at roughly 21 percent for a few hundred million years. This is notable because the atmosphere wasn't always 21 percent oxygen. Heck, for the longest time it was zero percent. But after fluctuating, it has settled at a steady level, which Lovelock would argue is regulated by life; this makes sense given that atmospheric oxygen is made by life in the first place. Interestingly, if the oxygen levels went down a few points, fire wouldn't burn, and if they went up a few, fires would burn out of control. In fact, fire may be one of the tricks Gaia uses to keep the oxygen level where it is.

What is so revolutionary about Lovelock's thinking is not just that the earth maintains the conditions for the survival of life, but that it links living systems and nonliving ones, blurring the lines between the two. The air, the oceans, and the land are not themselves living, but they change and morph with the life on our planet in a symbiotic dance of eons. The vast expanse of limestone rock upon which my native Austin, Texas, rests is a good example of this. Although my kids and I frequently find fossils in the limestone, which attests to its biological pedigree, the rock is also largely made up of carbon that Gaia pulled from the atmosphere eons ago and stashed away, presumably to help keep the oxygen levels of the atmosphere in check. And it isn't just limestone: Virtually all rock you see on the surface of the earth is biogenic in origin. Life has transformed, in one way or the other, almost every bit on this planet. Lovelock brings this point home, writing, "There is no clear distinction anywhere on the Earth's surface between living and nonliving matter. There is merely a hierarchy of intensity going from the 'material' environment of the rocks and the atmosphere to the living cells." Is any part of the earth untouched by life? Lovelock weighs in on this, too: "It may be that the core of our planet is unchanged as a result of life; but it would be unwise to assume it."

Although Lovelock, who died at 103 while I was writing this chapter, was a scientist with a laundry list of astonishing accomplishments, such as inventing the device that helped us identify the ozone hole, his theory wasn't well received by scientists when it came out. Its rejection of reductionist thinking in favor of systemic thinking made it seem a bit touchy-feely and unscientific. Lovelock also believed that the name itself was an impediment to it being treated seriously. It sure didn't trigger a groundswell of scientists naming their pet theories after pagan deities. As much as the name resonated with the public, it alienated scientists. However, "earth systems science," a similar view of the planet with a far more palatable name, seems to have warmed people to this way of thinking, and that, coupled with certain accurate predictions made by the Gaia hypothesis, seem to be helping this theory along as well. That said, it is still a minority viewpoint regarded by many as pseudoscience. Richard Dawkins, for instance, says in his book *The Extended Phenotype* that the hypothesis defies evolution. He writes, "The fatal flaw in Lovelock's hypothesis would have instantly occurred to him if he had wondered about the level of natural selection process which would be required in order to produce the Earth's supposed adaptations." Basically, he is saying that you would have needed tons of planets to have evolved something like the earth, and the galaxy would be full of dead planets that didn't make the cut. Even if you believe that, how would the successful planet have passed along its genes?

Another criticism leveled at the Gaia theory is that it is teleological in nature, that is, it implies an intentional direction to the whole process—an ultimate purpose of promoting life. Perhaps ultimately it is teleological, but that sort of phrasing can also be viewed as a linguistic convenience, the way one might say that natural selection "chooses" what will survive based on fitness. I find I have done the same thing in this chapter, using phrases like "Gaia wants" or "Gaia uses." Lovelock, for his part, pushed back on this criticism: "Nowhere in our writings do we express the idea that planetary self-regulation is purposeful, or involves foresight or planning by the biota." And yet, this strikes me as a carefully worded sentence.

Lovelock certainly regarded Gaia as a superorganism, using the term in the same way we are using it in this book. He wrote, "The entire range of living matter on Earth, from whales to viruses, and from oaks to algae, could be regarded as constituting a single living entity."

Few would dispute that the earth's various systems are, well, systems in the formal sense of the word. But is Gaia actually a superorganism? Referring back to our litmus test earlier, we see that Gaia does satisfy all of our conditions for superorganisms, as well as all of the conditions for life . . . except, it is alleged, one: It cannot reproduce, which Dawkins touches on above. But I differ in this objection. It definitely can reproduce, but we won't be ready to address that until section III.

In modern times, other planetary superorganisms have been postulated. One of these was offered by Kevin Kelly, mentioned earlier. He suggests that the world of technology that we have built, which becomes ever more integrated and connected, has coalesced into a superorganism he calls the Technium, which "has its own inherent agenda and urges, as does any large complex system, indeed, as does life itself." It is composed entirely of what he calls "things we have invented," and it has a "force that it exerts. That force is part cultural (influenced by and influencing of humans), but it's also partly non-human, partly indigenous to the physics of technology itself. That's the part that is scary and interesting." Kelly is right about that last part. For instance, the fact that a third of marriages begin with matches on dating sites shows us that the algorithms have started selectively breeding humans. But to what end? Kelly goes so far as to suggest that while we currently divide the living world into six kingdoms, such as plants, animals, and fungi, we might want to start considering the Technium to be the seventh kingdom of life.

It is tempting to ask which one of these theories is "right," but they are not mutually exclusive. When we view life in nested hierarchies instead of just single living creatures, we find that we inhabit a world of all manner of overlapping life-forms, and there may well be many superorganisms, with humans being a part in more than one. Gaia may well be a superorganism of all life, and Agora may also be one that emerges from the interaction of humans. Recall the opening line of this book—we are dealing with multiple creatures that coexist in the same exact matter.

CONCLUSION OF SECTION I

IN SECTION I, WE examined life. We started at the beginning, or at least as close to the beginning as our science and imagination can take us, and we explored its story—its long journey from simplicity to complexity. Superorganisms show up relatively late in that account and grow in complexity over time as well, adding layers of complexity on top of each other, ultimately reaching the size of the entire planet. Agora is one of these planetary superorganisms.

Section I was about that journey to ever more complex life, with new emergent capabilities forming at each level. We now understand what superorganisms are, and we are suitably impressed with their abilities. But so far it has been like a magic show. We've marveled at their amazing tricks, but now we are going to peer behind the curtain and see how the tricks are done. How do superorganisms self-create? How do they coexist in the same physical space as other creatures? How do they evolve at multiple levels at once? How do they acquire all the amazing capabilities that we have seen? How are they inexplicably so smart?

I don't know about you, but when I am shown how a magic trick is done, I'm always a little disappointed. They are usually quite clever but also pretty simple, and once I know the secret, I can never be wowed by the trick again. I'm going to show you, as best I can, how the superorganism's tricks are done.

While I don't know how life formed, nor do I know how consciousness comes about, I do think that by the end of section II I can show how all of the rest of the tricks are done. Consider that a spoiler alert.

I've said a few times that superorganisms acquire their emergent properties through the interaction of their parts. This is true but incomplete. It isn't the interaction of the parts that creates the superorganism but the interaction of specific forces that manifest through those parts. I'll elaborate on that shortly.

There are just six of these forces. Five of them are required for superorganisms, with the sixth, virtually unique to humans, being optional. That's it. You can understand the emergent properties of superorganisms by understanding these six forces and how they interact with one another through their manifestation in the parts of the superorganism.

The six forces are energy, information, communication, cognition, specialization, and, the optional one, technology. That's it. The bee dance, for example, is a combination of energy, information, communication, and cognition each expressed in a prescribed way. I wish the six forces spelled out some cool acronym like BREATH or something, but what we have is simply EICCST, which, regrettably, doesn't exactly roll off the tongue.

So what did I mean by *It's not the parts, but the forces that manifest through the parts*? Imagine a tug-of-war game. On each end of the rope are ten people pulling with all of their might. There is no specialization, nor is there necessarily communication, nor is information exchanged between the parts on each team, nor is any particular cognition going on. Everyone is just there pulling, so there is only energy. The two teams are basically two heaps of people, pulling in opposite directions. If you add another ten people to each team, or even a hundred, no emergent properties arise. They are still a heap.

Now imagine that those humans start playing American football instead. Everything is different then. There's lots of specialization: you have roles for lumbering big guys, speedy little guys, guys who can throw the ball far, and so forth. There's lots of communication going on in the huddle, with players exchanging information on which plays they should run next. Because of this, extensive cognition is occurring. And guess what? There are now emergent properties. The team can do things that no individual can, such as throw a ball fifty yards, catch it, and run roughshod over opposing players.

But here's the thing: it's the *same people* in both cases. It's the same parts. It isn't the mere interaction of those parts that makes the emergent capabilities, as if they were baking soda and vinegar, and 100 percent of the time they are combined you get a science-fair volcano eruption. Rather it's the interaction of the forces that they manifest that makes the magic. The parts, be they people or bees or whatever, are merely the vessels by which the interactions are brought about.

You could, after all, teach two humans to communicate through the waggle dance. They could convey to each other the location of the good flowers and how to get to them. It's not the vessel that matters but the forces that they embody. However, humans would do an incredibly poor job of waggle dancing and would totally lose a dance-off to the bees. Why? Because the bees have evolved to embody those forces in a specific manner that is uniquely suited to their particular environment. You could also teach two robots to communicate via the waggle dance. They are even less suited to it than people, but you could theoretically do it. This matters because it demonstrates that the parts needn't necessarily even be biological. But would a hive of robot bees be alive? It would check all the boxes for a living creature, except one: it, too, would run up against its inability to reproduce, at least with our current technology.

The six forces—energy, information, communication, cognition, specialization, and, finally, technology—are concepts that we encounter on almost a daily basis. But we generally use them in their colloquial sense, as in "I don't have the energy to do that" or "Umm, that's a little too much information." In everyday life, we don't have any real need to understand abstractly what energy is, or information, or any of the rest. But for our purposes here, we do. Those six forces account for *all* superorganisms, all the way from beehives to Gaia, and everything in between. So presumably they must account for Agora as well. We've already seen all of these forces at play in the superorganisms we have examined so far, but just anecdotally. We need to understand them systematically, even philosophically.

So that brings us to section II, "Forces." We'll spend a chapter understanding each of them at its core so we can see in section III how they apply to Agora—how a city is a hive and can encode information, how walking down the street is form of cognition, and how specialization can be solely cognitive with no physical expression. We'll see how Agora thinks in the same fundamental way that a beehive or a human thinks.

SECTION II
FORCES

THE MAMMOTH HUNTERS

IMAGINE YOU LIVED A few hundred thousand years ago, in the earliest days of our species. There was virtually no technology, no agriculture, no cities, no language, not even any clothes to wear. Although humans were anatomically modern in the sense that skeletal remains from that time appear no different from those of today, we were clearly not yet human in any other sense of the word.

What would daily life have been like? We don't know, of course, but we can make some reasonable guesses. We probably skulked around, staying out of sight in a world we were not yet masters of, always on the lookout for danger. The sound of a twig snapping from an unknown source must have been cause for worry and caution, as we were prey as well as predator. As hunter-gatherers, we would have wandered around picking berries and rolling over logs looking for grubworms to munch on.

Let's say that your group happened to spot a mammoth. Would they have tried to harvest the animal? It would be a tempting target given that a mammoth had roughly enough calories to sustain the entire band—maybe a hundred people—for a month. But in spite of that, we would not have tried to take down such a beast. We are pretty confident about this because we have no archeological evidence for humans of that era having defeated a mammoth, but we have literally thousands of examples of it from more recent history. Had they been bold enough to try, the mammoth would have likely found them more annoying than frightening, and it would have just rolled its eyes, or whatever the mammoth equivalent was, while deciding who to trample first.

Now, fast-forward thousands of years to a time when humans do have language, technology, and all the rest. Let's say it is the Upper Paleolithic, about twenty thousand years ago, at the height of the last ice age. We are, by this point, indisputably modern humans in all senses of the phrase—and quite proficient mammoth hunters as well. What changed? Let's have a look:

Imagine you are in a band of five Paleolithic hunters whose mouths are all watering for a nice, thick mammoth steak, and as luck would have it, you spot a great big mammoth off in the distance. Surveying the land and the tools at your disposal, you all concoct a plan: One of you will sneak around to the other side of the mammoth, another will climb a tree and go out on a branch above the mammoth, and so forth. At the appointed time, you would attack in tandem and execute the plan. As it unfolded, you would naturally shout information and commands to one another. "He sees you!" "Now, attack his back legs."

The mammoth is no longer facing a mere annoyance, but, from its viewpoint, an unimaginably hideous monster: It has ten eyes and ears, ten arms and legs, five heads, and, here's the kicker, one mind. It acts as a single entity, a solitary being, at least for that brief moment.

To be clear, this band of hunters is not a superorganism. While we do see emergent capabilities—the band can do things no individual can do—the members of the hunting party will go their separate ways after the hunt, so we aren't really dealing with an indivisible creature. But if you recall from earlier, our journey from single-celled creatures to multicellular ones may have begun this same way: at first, the parts came together temporarily to solve a particular problem, but over time, their interdependence and specialization reached a point where emergent properties were brought about, and then they ceased being solely individuals and became something more as well.

That's what's going on with the mammoth hunters. And if you look carefully, you can see something resembling a superorganism of humans starting to form in outline. And, just as single-celled creatures did, over time, humans gradually became more specialized and more dependent on one another, growing less able to survive alone, and finally crossing a threshold to become a bona fide superorganism.

When did this tipping point occur? When did we cross that threshold? We can answer this question with some confidence. It happened with the formation of cities. Before that, our six forces weren't around in enough abundance to get the ball rolling. As my grandma always used to say, "A dash of information with a pinch of communication does not a superorganism make." The city provided the quantity and concentration of these forces necessary for the superorganism to form. The archeological record bears testimony to this. Before the city, while our technology did gradually improve, our mode of living didn't change all that much. Over time, we built somewhat better shelters, became slightly more successful hunters, and incrementally improved our quality of life, but all of this was by degrees. When we got cities, however, boom! Everything suddenly changed. Agriculture led to surpluses that allowed for specialization, which created emergent properties. No band of hunter-gatherers, even among the hundreds that lived to the present day, ever entered the Bronze Age, or developed writing, or any of the thousands of other capabilities that sprang into being when we moved to cities.

As hinted at in the superorganism chapter, we think bees followed a similar path to become a superorganism. Whether because the population of solitary bees grew or climate change shrank their habitat, they seem to have started living in closer proximity with one another. At that point, they perhaps began to slightly specialize, with one bee managing the brood and another gathering food. That strategy—specialization—is so powerful and transformative that it gave them an advantage: the more social bees outperformed the solitary ones, at least in some species in some areas. Their relative success led to their thriving over their more aloof counterparts. Over time they would double down on that strategy, again and again, until they crossed the threshold and emerged as a superorganism.

Even though our Ice Age mammoth hunters aren't a superorganism yet, I will be referring back to them regularly because it is so easy to see all six of our factors at play here, interacting with one another to create emergent abilities.

ENERGY

OUR MAMMOTH HUNTERS NEED energy to accomplish anything. Without it, they would be as lifeless as rocks. I list this force first for this reason.

What is energy? It is defined almost everywhere as the capacity to do work. This isn't a great definition, because you may have noticed that it doesn't say what it actually *is*, only what is does.

Energy exists in two broad buckets: potential and kinetic. There are other types of energy that go into each bucket, such as chemical, nuclear, heat, light, magnetic, elastic, gravitational, electrical, and several more. All those forms of energy are measured in units called joules, a fact that allows us to draw equivalencies between them.

But even with that additional detail, we haven't addressed what energy actually is. That's because it is not a thing, like water. It's more like a relationship than anything else. A rock resting on the ground has neither potential nor kinetic energy. If you pick it up and hold it in the air, it now has potential energy. But what changed? Its weight is the same; it doesn't react differently to a magnet; it's the same rock. We just say it has a certain amount of energy. If you drop it, then its potential is converted to kinetic energy, but again, the rock isn't any different. When it hits the ground, the energy is gone.

Gone? Where? It went off to do other work. That's what energy does. That's its "thing." It is always conserved, meaning there is no more or less of it now than there was a million years ago. It's like double-entry bookkeeping: Every credit has to have an offsetting debit, and those two columns must always be equal. When you lifted the rock, you didn't make any energy, you just transferred energy from your muscles to the rock. When the rock fell, it "released" energy back into the universe. Crazy, isn't it? And to top it all off, we don't know *why* energy is conserved. It just seems to be the way things are.

As we discussed in the chapter "What Is Life?" there is a relentless march in the universe from order to entropy. This doesn't mean that the energy is lost, just that eventually it will become evenly dissipated throughout the universe, thus unable to do any work. The only way you can stem the tide of entropy is by using energy to create temporary pockets of order. We are lucky in that we live on a planet that has energy pouring down on it from a nearby fusion reactor, so much so that we can create huge amounts of order, a.k.a. life.

The earliest life most likely got energy through chemical reactions and stored that energy in ATP. We know this, as mentioned earlier, because all life runs on ATP, so it must have been what powered LUCA, our Last Universal Common Ancestor. ATP, you might recall, is like little rechargeable batteries in the cell that are used to run all life. Each cell uses millions of ATP molecules every second.

Sometime later, an enterprising life-form learned how to turn sunlight into usable energy. We call that photosynthesis. Much later, other creatures came along and started eating those creatures, absorbing their energy. This is predation.

Creatures show a remarkable versatility in how they take energy from their environment and use it for power. Through its yellow stripes, the oriental hornet absorbs sunlight that it can put to use, and we recently discovered a fungus near the Chernobyl nuclear power plant that can convert the plentiful gamma radiation to chemical energy.

That brings us to our mammoth hunters—and ourselves. Our bodies use 100 watts of power to do all that they do. That's a straightforward calculation based on eating 2,000 calories a day, which works out to about 80 calories an hour, which works out to about 100 watts. Watts, by the way, are a measure of power, not of energy. Think of them like miles per hour with a car, except they are joules per second. A 100-watt light bulb requires a constant 100 watts to run. A battery doesn't "hold" 100 watts—that's a nonsensical statement. It can, however, hold 100 watt hours, meaning that it could generate 100 watts for one hour or 50 watts for two hours.

The difference between what your body uses while you are asleep versus when you go work in the field is quite substantial. While you are sleeping, your body requires 50 watts to keep everything running, but hard physical labor— the kind humans can only muster for brief periods—can use several times that. But for most of us that averages out to about 100 watts of power.

That's it. If you were dropped on a desert island, you would quickly feel the limits of what 100 watts can do. There are only two ways around this limit. The first is by figuring out how to do more with our 100 watts. We call that technology, and we will dive into that in a few chapters. The other way is to harness other sources of energy and put them under our control. For humans, this began with our use of oxen for plowing roughly 8,000 years ago. An ox is a wonderful engine capable of turning 4,000 calories of grass a day into 200 watts of power with which to pull a plow, so using just one triples your energy budget, minus, of course, the energy you have to spend tending to the ox.

From that time until just a few centuries ago, the only sources of energy humans could use for work came from living creatures. When we started using machines powered by fossil fuels, the energy we had access to skyrocketed.

Why we have fossil fuels at all is a fascinating topic. About half a billion years ago, moss from the ocean crept onto the land, liked the weather, and decided to live there. Eventually it turned into lichens, giant ferns, and three-story-tall mushrooms. Trees came much later, and are in fact younger than sharks. There was a pivotal period of tens of millions of years when all those plants would grow, mature, and die, but not decay when they fell over. That last bit is the important part: no microbes had yet evolved to feed on them, that is, to break them down. So more plants grew up on top of those, until they, too, fell over dead. On and on. Eventually, microbes evolved to break down dead plants, but by then, those eons of accumulated biomass had become compressed under its own weight such that each fifteen feet of dead matter became one foot of coal. A forty-foot-thick vein therefore was once six hundred feet of dead mushrooms and ferns piled upon themselves. This is why plant fossils are found in coal.

We later learned how to take the energy from those dead plants and put it to use to achieve our goals. Unfortunately, this also releases all that stored carbon back into the atmosphere. But our consumption of this ancient energy is what we owe much of our success as a species to, as well as virtually all of our prosperity. If you live in the West, then the total energy consumption divided by the population shows that on average we each use a steady 10,000 watts of power, a hundred times more than your body runs on. Look around at the highways and skyscrapers and all the rest, and you will see how we put that energy to work.

Incidentally, there are multiple paths away from fossil fuels. Fortunately, energy is incredibly abundant and several other sources are available. For instance, the interior of the earth has not significantly cooled since the planet was formed and is still as hot as the surface of the sun. This surprising fact is due to two things. First, the crust of the earth is a fine insulator, and second, a great deal of radioactive decay is going on inside our planet, which generates heat. In fact, the core may even be getting hotter each year, and even if not, estimates are that the warmth of our planet's interior will outlive the sun by many billions of years. This same dynamic is going on in some of the outer planets and moons, bodies so far away that they get virtually no warmth from the sun but may still even harbor life that would be powered by warmth from within. On Earth we might soon be able to tap this unlimited source of geo-thermal clean energy, given that just six miles down—a depth we can already drill to—there is fifty thousand times more energy available than all the oil and natural gas reserves on the planet.

However, the difficulty of getting rid of fossil fuels is driven by their abun-dance, low cost, and, perhaps most importantly, their energy density. Picture how much food it takes to feed you for a day. That's the same amount of energy as in a teaspoon of crude oil. It's hard to beat that, given that it is about a hun-dred times more energy dense than honey. By the way, the reason that sugary drinks have so much appeal is that your body sees them as a high-energy source, and likewise, artificial sweeteners taste so awful because your taste senses are so exquisitely sensitive that they know when they are being tricked.

While in the West we use a constant 10,000 watts per person, worldwide the average is 2,500. If you divide what humans spend on energy across the world's population, you will find that those 2,500 watts cost about three dollars a day. Isn't that astonishing? All that power is just three dollars a day. Unfortunately, many countries must make do with a small fraction of that. In 2013, Ellen Johnson Sirleaf, the president of Liberia, pointed out that the electrical load used by the Dallas Cowboys' stadium during a home game was triple what her country of four million people was able to generate. There are several countries even worse off than this, where people manage on just 400 watts or so of power. That's basically two oxen as their entire "bonus" energy, so they are living a preindustrial life. This effectively locks those countries into perpetual poverty since there is no better correlate to standard of living than energy consumption. Why? Because the more energy you consume, the more "work" is being done, and thus the more productive a person becomes. A person hoeing a field can make very little income, the same person with an ox can make much more, and give that person a tractor and they'll become comparatively rich. And in addition to the dollars-and-cents aspect, more energy consumption equates to better quality of life as people are able to use that energy to "buy back" time they spend using their bodies as little more than machines.

Here's a question for you: What if our energy consumption went down by half? All over the world. In the developed West, that would mean going from 10,000 watts of constant power to 5,000. It might seem like that would be little more than an inconvenience. After all, that's still twice the average energy consumption of the planet. Ahh, but this isn't right. In reality, virtually everything would collapse. We have built a society so complex and interconnected that it requires 10,000 watts to run. Cut out half our power, and we aren't half as wealthy; it's more like having only one AA battery for a gadget that needs two. It's not going to run at all.

That's the first of our six forces. Energy is important to our story because it is the only thing that can keep chaos at bay, and it's also the only thing that can animate the parts of a superorganism to produce emergent properties and bring about the next higher level of complexity.

But this is worth keeping in mind: Put aside the energy that humans are able to deploy to power their machines, and let's think just for a moment about biological energy. Cells use ATP as their power source. Bees, in turn, are powered by their cells, so, by extension they run on ATP as well. Thus the hive isn't really powered by bees, as I indicated earlier, but by the bee's cells. The bees are just a vessel by which cellular energy is put to use at the hive level. This goes all the way up. Gaia, for instance, is ultimately powered by cells as well. A whole, whole lot of them. (All of this is not quite absolute, though. The sunlight that shines down on the hive helps keep it warm, so the hive in a sense is solar powered. But this is barely even incidental in the grand scheme of things.) The point is that ultimately Agora is powered not by us but by our cells. But you can't really go straight from cells to Agora. What do I mean by this?

A car's drivetrain is ultimately made of steel. But you can't pour steel into a mold to make a drivetrain. You must first make gears, then combine gears to make a transmission, then combine the transmission with other complex parts, such as the engine and differential, to make the drivetrain. But, in the end, the drivetrain is just made of steel. Agora is thus ultimately made of and powered by cells, and we are simply an intermediate part whose function is to step up complexity one more notch.

But again, we are a special case. Because we use energy sources other than ATP, such as petroleum, we can create far more complexity than we could with just our cells. And while from a planetary standpoint we use twenty-five times more energy than our bodies generate, we amplify that energy with technology such that we create thousands of times more complexity than we could with just our cells and biology. This makes Agora unfathomably powerful.

INFORMATION

OUR NEXT FORCE IS information. Our mammoth hunters had a good deal of it: They knew the mammoth's location, as well as its general behaviors. They had weapons, and they knew what those could and could not do. Most importantly, they each had a lifetime of learning and the information they had accumulated.

We want to understand what information is in an abstract sense so that we can apply it to a range of organisms, from cells to plants to us. First, let's clarify the difference between data and information. Data is just facts, while information is how we arrange and interpret those facts. This difference is easy to illustrate. Say you hit your thumb with a hammer. You think, "That hurt." That's a piece of data. If you do it again, you think, "That hurt, too." Eventually, you learn some information: "Hitting your thumb with a hammer always hurts." At that point, the data becomes unnecessary and can be thrown away, as long as the information is kept. Our brains are good at doing exactly this. In other words, while you will remember that it hurts to hit your thumb with a hammer, you probably won't be able to recall any specific time you did it.

That's one way to obtain information: via sensory experience combined with memory. The problem with this method is that this requires sophisticated neural hardware that comparatively few life-forms have. To complicate matters, data and the encoded form of that data don't look anything alike. Genes that give you blue eyes aren't blue. That means that creatures must have a method to translate an experience into a piece of data and then encode that data in memory. They also must have a way to access that data later and unencode it.

That's a tall order for, say, a fern or a beehive. So how do creatures with limited cognitive abilities transform data into knowledge? Often it is just hard-wired in their DNA. Individual creatures that don't have some key piece of information written there die, while those that do have it thrive and reproduce, preserving and passing down the information.

But that's the long game. How does a creature with limited cognition acquire data and transform it into information within a lifetime? How does it encode and access that information? The answer is that the act of living combined with the passage of time can itself encode information. Creatures evolve specifically to be able to access that information. Let me unpack that a bit.

Think about Sherlock Holmes and how he deduces things. He could walk into your empty office—if he weren't fictional—and start telling your life story. He would see a hundred things that would allow him to make inferences about you, such as the wear pattern on your rug, the height your chair is set to, and the way you place your pen on your desk. Your daily living encoded that information.

Think about desire paths. Those are the walkways that appear in places such as universities that are created when multiple people take the same short-cut between, say, two buildings. Angela cuts diagonally across a patch of grass to get someplace a bit quicker, then Bob does the exact same thing, then a hundred more people do it, at which point there is a well-defined path. This path is a piece of information about the preferred way to get from one point to another. It has been encoded not with ones and zeros on magnetic media but in compacted dirt. The data—each instance when someone took the shortcut—is long forgotten, but the information remains. This is how superorganisms can encode information as well. The actions of individual ants or bees over time result in the mound or hive having a certain layout that is optimal for its specific situation. Each way that two ant mounds or beehives vary from each other—and there might be thousands of tiny ones—is a piece of encoded contextual information.

Sometimes, data is passively stored inside the creature. The rings in a tree tell a complex story. Not just the age of the tree, but when it encountered fire or a drought, even if those happened centuries ago, and likewise, the number of layers of earwax in a whale tells its age as well. Other times, information is stored by non-biological processes. Visitors to the Grand Canyon can see seventy million years of encoded information in its various strata.

This is how information can be encoded without any intelligence at all. When we come to the chapters on cities, we will see how they also encode vast amounts of information in these exact ways.

But how can a creature access such information? Doesn't that require a brain? No, this can be done passively as well. There is a stretch of highway in Lancaster, California, called the Musical Road. Grooves are set in it in such a way that if a car drives over them at 55 mph, it plays the *William Tell* overture. The car isn't intelligent nor is the road, and yet the car can access the information on the road. Likewise, in our example above about the ant mounds and beehives, insects access that information with virtually no cognition by simply interacting with the mound or hive.

Storing and retrieving information is so key to survival that nature has evolved thousands of techniques to do so. For instance, Venus flytraps can count. To make sure that their traps don't get sprung with a single raindrop, they require three touches to a trap within a certain short period before they expend the energy needed to close. Thus, they must be able to count the touches as well as store the time interval between them. We actually know how this is done. When a trap is touched, some calcium is released by the plant, which slowly—over seconds—dissipates. Subsequent touches release more calcium, and if it hits a certain threshold, the trap snaps shut. Since the calcium is constantly dissipating, the sequential touches have to be within a certain time window to hit the threshold. This capability is clearly encoded in DNA somewhere, but again, information is being acted upon with hardly any cognition.

Similarly, *Mimosa pudica,* commonly called sensitive plants, seem to be able to learn. They, too, have a mechanism that closes their leaves, which is activated if they are dropped. But if they are repeatedly dropped, they stop closing their leaves, "realizing" that it is not a threatening situation. One might suppose they can't close anymore because they are just fatigued. But researchers return to the same plants a few days later, drop them, and find they still don't close. The plants have "learned" that this is not a situation that calls for it. I put "realizing" and "learned" in quotes because those may not be the best words to describe what's going on, but we don't have any better ones. While we don't know how they remember, it might be something like how the flytrap counts. What's interesting is that information is being somehow encoded and later accessed. It isn't inconceivable that after some number of generations, offspring could evolve to stay open when dropped. That would bridge the world of short-term memory with the long-term one of DNA, which has always been a bit of a mystery.

That's three ways that creatures can store and retrieve information: through DNA, through sensory experience coupled with memory, and through incidental or passive encoding. Are there others? Just one: we get information from other creatures, and that's called learning, which is done via communication, another of our six forces, which we cover in the next chapter.

We are going to see all four of these information storage and retrieval methods active in Agora, but at an exponentially increased scale. We'll see that humans, in addition to passively writing data through the activity of daily life, are able to deliberately encode data for anticipated needs. This requires an understanding that there is a future to plan for, a capacity unique to humans.

I want to spend the remainder of this chapter talking about human methods for information storage and retrieval in particular since we are a bit of a special case.

Let's return to the topic of DNA. Its two key attributes are that it encodes information and it has a mechanism to reproduce itself. Without the latter, information is good for only one generation. It can't persist. For billions of years, from the very first creatures up until about seventy-five thousand years ago, DNA was the only place to store information long-term. It worked, to be sure, but boy did it take a long time to write down just one thing. Many thousands of years, in fact. Thus it would take an animal eons to evolve an instinctual aversion to a certain berry that would make it sick.

Interesting, isn't it? DNA, as noted earlier, is just a book of letters. It isn't alive or anything. The animal only knows not to eat the berry because the

DNA created the animal in such a way as to avoid it. Consider the koala, which is, shall we say, not a mental giant. It likes to eat eucalyptus leaves, a preference that is encoded in its DNA. However, the information that is stored there is so meager that if you *handed* a eucalyptus leaf to a koala, it wouldn't know to eat it. If you put the leaf on a table in front of it, the koala wouldn't recognize it as food. It only knows how to recognize it growing on the tree. That's all that is written into the DNA.

About seventy-five thousand years ago, humans got language. Something happened to us that seems to have given us all of our other unique abilities as well—our creativity, knowledge of the future and the past, music, art, technology. Everything that makes us special seemingly all came to us at once. This event goes by a number of names, including the Great Leap Forward, the Cognitive Revolution, the Symbolic Thinking Revolution, the Creative Revolution, the Upper Paleolithic Revolution, and the Human Revolution. In my book *Stories, Dice, and Rocks That Think*, I call it The Awakening and spend much ink there sorting through exactly what happened.

Surprisingly, the most important aspect of language is not communication but rather as the instrument of thought. We think using language, and our brains are able to later recall those thoughts. The big deal, therefore, with language was that for the first time ever, a species could write information to its brain in addition to its DNA. But that's only half the story. Sure, you can write thoughts to your brain using language, but how do those thoughts spread to others and get passed down to other generations? That required the invention of speech, our method for transferring our internal language to others. Just as the information in DNA is encoded and spread via biological heredity, information could be, for the first time, preserved and spread another way, through the transfer of thoughts via speech. The resemblance between these two methods is uncanny. If you ate a berry and got sick, you could tell people, "Don't eat the little purple berries," and then that mutation—this new information—would spread throughout the population at lightning speed. Forget about thousands of years to write something to DNA and then thousands more for it to propagate throughout the species. Now it could all happen in months, or years at most. This new mechanism for storing information was a big upgrade from

DNA since it had so much more storage capacity and operated so much faster. This giant leap forward accelerated our evolution a million-fold, enabling us to become the preeminent life-form on the planet.

There things stood until something else profound happened. Instead of just having DNA and our brains to store information, we invented writing. Who saw that coming? We could now store information externally, and it could be copied—just like DNA—and it could in theory last forever. You could write a book of herbal lore that explained why you shouldn't eat the purple berries, and that book could also go into great detail about how they can be boiled into a tea that is good for treating lumbago. That book is a new gene, and that gene can be copied and spread around the world. This is why I maintain that paper is the most important invention of all time, because it enabled us to externalize our genome.

Our genome used to be solely contained within our DNA. With speech, information could be stored in the brain as well, so our genome was effectively what was in our DNA plus what was in our brain. With writing, our genome became all of that plus whatever books were in our library. Today, that means everything that is on the internet. That's humanity's collective genome now. It is unfathomably large, but just as our DNA has a great deal of "junk," so does this new expansive digital genome.

The real power in all of this was that information could accumulate over time. That's really our secret sauce as a species. Each generation of humans begins with the virtual genome of the world it was born into—that is, everything that was written down that has survived over the last five thousand years. That corpus continues to grow throughout our lifetimes such that our children are born with a larger, and hopefully better, virtual genome than we were. That's what we call progress.

Consider this story: In 1864, five men we call the Grafton castaways used their combined smarts to survive eighteen months shipwrecked in New Zealand's Auckland Islands. With only two months' provisions, they survived on seal meat, birds, and fish. They built a permanent stone-and-timber cabin, complete with sleeping stretchers, a fireplace, a dining table, and a writing desk. One man constructed a forge to work metal, while others tanned leather, made soap, and even brewed beer. Eventually, using the forge and bellows to craft new tools, the men were able to construct a small dinghy by which they made their escape.

I have two comments to make about this. First, wow. Second, consider how that all happened. They didn't invent all that stuff. It was in their *virtual* genome, as it was part of the knowledge they were born into. Undoubtedly, they adapted that to their conditions, but they could do all of that not because they were so smart, but because five thousand years of accumulated knowledge made its way down through history to them. This is just like any other heritable gene. The difference is the scope and speed of the process.

Humans are also exceptional, although by no means unique, in terms of our range of senses, which we use to acquire most of our information. About a century ago, Edwin Hubble, the famed astronomer, said, "Equipped with his five senses, man explores the universe around him and calls the adventure Science." And what senses we have. A nanometer is a billionth of a meter. A sheet of paper is about one hundred thousand nanometers thick. Guess how sensitive your fingers are in terms of the smallest things they can detect? Just twelve nanometers. Professor Mark Rutland of the KTH Royal Institute of Technology in Stockholm put this into perspective thus: "This means that, if your finger was the size of the Earth, you could feel the difference between houses from cars . . . We discovered that a human being can feel a bump corresponding to the size of a very large molecule." And yet these are the same digits you can use to lift up to twenty pounds or engage in a thumb war. Eyes are similarly impressive: If it were truly pitch-black outside, you could see the flame from a match strike on a mountain thirty miles away. And yet they are the same eyes that don't get overwhelmed by IMAX movies. Interestingly, a quarter of people who grew up watching black-and-white television dream in those hues, while the younger generation no longer does, showing just a bit of how our senses influence our view of reality, and vice versa.

And then there is hearing, which influences our perception of our other senses. As Seth Horowitz, author of *The Universal Sense: How Hearing Shapes the Mind*, puts it, "You hear anywhere from 20 to 100 times faster than you see, so that everything that you perceive with your ears is coloring every other perception you have, and every conscious thought you have." He adds that sound "gets in so fast that it modifies all the other input."

Our sense of smell is an odd one. We lack a vocabulary to describe most smells, so we can only say things smell like other things. This gets us by, and we hardly notice the limitation. There is a widespread belief that our sense of smell is poor compared with not just dogs but also apes, who seem far more in touch with their olfactory senses than us. Why would this be? One theory I find pretty believable—and a little amusing—is that since the development of cities and agriculture, we tended to start sleeping jam-packed in small structures like tents or shelters. For a species that hadn't yet invented the bathtub, this must have been rather odoriferous, so we coped by evolving a weaker sense of smell.

It is worth noting that where we have no senses we have a good deal of ignorance. We can't see infrared nor ultraviolet wavelengths, so we don't know how the world looks bathed in those lights. Oh, wait, yes we do. We have built machines that can detect them and those devices function as additional sensory organs for us. We might think that hawks have better vision than we do, but really, they don't. As good as their eyes may be, they don't hold a candle to the microscopes and telescopes through which we perceive the world. Because of technology, our range of senses is greatly expanded and thus our access to information as well.

That's our second force, information. Superorganisms need a mechanism to acquire and store new information. They have multiple ways to do this requiring almost no cognition. Without learning and memory, they can't adapt to change. Humans have exclusively used DNA for most of our history, but now the complete accumulation of all of our knowledge—in our brains and our books—is our virtual genome, which would by extension be Agora's DNA.

COMMUNICATION

OUR NEXT FORCE IS communication. Our mammoth hunters communicated among themselves as they made their plan, and the successful execution of the plan required more communication. We should be able to see in Agora a vast amount of peer-to-peer (not top-down) communication.

It is said that all living things communicate in one way or another. "All" is a pretty high bar, but certainly virtually all do. The range of ways they do so is enormous. Bioluminescence is perhaps the most common method, but others use percussion, electrical impulses, odors, sounds, and, of course, the dance of our honeybees. Plants get in on it as well, communicating through their roots or via chemicals released into the air. The cells in your body communicate with one another in a number of ways, such as through the use of hormones. The difference between hormones and pheromones, by the way, is that the former are internal to a body and the latter are external. For our purposes, that distinction isn't all that meaningful because we are looking at creatures whose bodies are distributed. When ants communicate by laying down pheromones, it isn't really any different from when our cells communicate using hormones. All of these methods of communication are of course instinctual. No one teaches a firefly how to flash nor what the signals mean.

Communication can be indirect, or even unintentional, the way that the parts of a superorganism might take cues from other parts, a sort of mash-up of communal decision-making, peer pressure, and mentoring. The parts can learn basic things from one another. Older honeybees—and we are talking just a few days older—do the waggle dance more accurately than the younger ones. Additionally, experimenters have found that bumblebees that watch another bee through a window figure out how to get at some delectable sugar water will immediately use the same trick when presented with the same challenge. The cells in your body also keep an eye on their neighbors and adjust their behavior accordingly. That's why an embryo begins as a single spherical cell but ends up with a head on one end and toes on the other. And of course, that's what we do as parts of Agora: we act autonomously but also with an eye on the actions of our neighbors and coworkers, which informs our own actions.

Another form of communication is language, which is unique to humans. As Bertrand Russell put it, "A dog cannot relate his autobiography. However eloquently he may bark, he cannot tell you that his parents were honest though poor." Real language, like what we have, has four characteristics that together are completely unmatched in the plant and animal world. First, language is made up of symbols. Animals do this part extensively. If a vervet monkey sees a snake and sounds an alarm, the sound is not an imitation of the hissing of the snake, but rather a specific call that just means there's a snake. Second, language is multilevel. For us, that means sounds make words, and those words, in turn, make sentences. This maximizes the amount of meaning that can be expressed with the fewest number of unique sounds. Third, language is productive, meaning entirely new ideas can be expressed simply by reordering the words. This means that while we have a finite number of words, an infinite number of ideas can be expressed with them. Finally, language has displacement, that is, the ability to talk about things in other places and other times.

Plenty of animals have advanced communication systems that satisfy one or even a couple of these criteria, but when it comes to language, we are so different from every other creature on this planet that we may as well be aliens living among them. Human language is incomprehensibly more powerful than any other biological communication system in the world. There aren't any creatures that have, say, 30 percent of our language ability, or even 3 percent or even 0.3 percent. There's simply nothing else like it. Its power is that it facilitates the

exchange of information, as well as promoting specialization and collaboration.
Think about how hard it would be for dogs to design a Saturn V rocket and go
to the moon just using barks.

I say all this to point out that social insects form superorganisms with communication that amounts to just a few words. There are no new words ever, and
the words cannot be combined in different forms to mean different things. And
yet, think of the amazing things they are able to do. The nuance possible with
our language gives Agora unfathomable power because if you take the gains in
information exchange that spoken language enables and multiply that across
the several hundred million conversations going on in the world twenty-four
hours a day, you get an incalculable—literally—amount of information flying around. That sure looks like a giant brain to me, with billions of neurons
chatting away all day long. As a general rule, more communication is better in
a superorganism. Bees and ants communicate continuously, as do the cells in
your body.

The modern world has built communication tools that enable ever more
people to communicate with each other easier and faster, compounding Agora's
power. A hundred years ago, a three-minute phone call from San Francisco to
New York cost $20, or something north of $1,000 in our modern money. Now,
we may get dozens or even hundreds of emails a day and spend hours communicating. This is more powerful than it may seem at first glance. Nikola Tesla
wrote an incredibly prescient piece about the future that you and I reside in. In
fact, it is so prescient, I always regarded it as apocryphal until I finally located
a copy of the January 30, 1930, issue of *Collier's* magazine in which it appears.
Tesla wrote:

> When wireless is perfectly applied the whole earth will be converted
> into a huge brain, which in fact it is, all things being particles of a
> real and rhythmic whole. We shall be able to communicate with one
> another instantly, irrespective of distance. Not only this, but through
> television and telephony we shall see and hear one another as perfectly
> as though we were face to face, despite intervening distances of thou-
> sands of miles; and the instruments through which we shall be able to

do this will be amazingly simple compared with our present telephone. A man will be able to carry one in his vest pocket.

Pretty much the only thing he got wrong was that men in the future would still be wearing vests.

But human communication isn't limited to just conversations between people. In fact, that's the smallest portion of it. We have built a vast communication network of machines that exchange orders of magnitude of information billions of times faster than speech can. To accomplish this, we created entirely new languages suitable for machines that promote even more efficient exchange of information. And while those machines aren't directly part of Agora, the abilities they give us filter up to Agora through us. Thus, our interconnected world of computers is a superpower for Agora.

In our story of the mammoth hunters, I contrast the pre-speech hunters with the ones who have language. The former just got trampled, while the latter had a huge feast. This was the ultimate enabling technology, the one that divides all human history into two parts: BL and AL—before spoken language and after. When did we make that monumental shift? While this is a contentious question, I believe the bulk of the evidence supports that it happened quite recently, between thirty thousand and eighty thousand years ago. Further, I believe the capacity for it happened to just one person, one time, on one day in history, and that person, who had an extraordinarily fortuitous mutation, is the ancestor of us all.

Whatever mutation gave us language bestowed a bunch of other powers as well, including music, art, and knowledge of the future and past. Although that seems like a whole lot of speculation, the archeological record sure seems to support this "instant transformation" hypothesis. Consider cave art. Our earliest representative cave art doesn't consist of stick figures but beautiful, nuanced depictions of animals. When we excavate some of our earliest caves with such drawings, we discover musical instruments for the first time, exactly concurrent with the emergence of art.

Further, because all human languages have those four properties described earlier, it is likely that they all came from a single "mother tongue." In other words, we developed language just once. The presence of cognates, that is, words that mean the same thing in multiple languages, supports this idea. The Basque people of Spain are said to have a language of such great antiquity that they believe it to be the original language. I have no particular opinion on this, but Basque is unquestionably a linguistic mystery. It is certainly ancient: the words for knife and hammer seem to be derived from the word for rock.

A superorganism can be powered by any form of communication, but language is exponentially more powerful because of its versatility as well as its information density. Additionally, language is how we communicate with ourselves, that is, it is the stuff of human thought. Not only can your dog not tell you that his parents were honest though poor, but the dog can't even think that. Language, at least internal language, could well be a requirement for consciousness. This is not purely speculation. In her 1908 book, *The World I Live In*, Helen Keller wrote about her life before her teacher came, that is, before she had language:

> I lived in a world that was a no-world. I cannot hope to describe adequately that unconscious, yet conscious time of nothingness . . . Since I had no power of thought, I did not compare one mental state with another . . . When I learned the meaning of "I" and "me" and found that I was something, I began to think. Then consciousness first existed for me . . . It was the awakening of my soul that first rendered my senses their value, their cognizance of objects, names, qualities, and properties. Thought made me conscious of love, joy, and all the emotions. I was eager to know, then to understand, afterward to reflect on what I knew and understood.

That's our third force, communication. Communication is required for collaboration, and that in turn is required for a superorganism. The parts of a superorganism frequently must coordinate their actions, which must be done through communication of some kind. Human language is a special kind of communication evolved to accommodate the extraordinary level of information sharing that we do, and in turn, we share more information because of spoken language. But even it is low bandwidth compared with the machines we build. It would take you fifty hours to read *War and Peace* out loud, but you can download all that text in a few seconds. We created these communicating machines and developed new languages suited to them in order to satisfy our wish to share ever more information, and there is no hint that we are reaching any kind of limit in either our capability or desire to share ever more.

You may be seeing a pattern develop in these last few chapters. With regards to energy, humans use exponentially more than any other creature. We encode and access exponentially more information, and, as we have just seen, we communicate exponentially more than anything else. When those are compounded on one another you get something that is not just exponentially larger but astronomically so. Perhaps you can start to see how a superorganism like Agora can emerge, and while Agora is the same *sort* of thing as a beehive, that is akin to saying that the Wright Flyer and a Boeing 747 are the same sort of thing. And we haven't even gotten to the other three forces in which humans are also vastly beyond anything else.

COGNITION

OUR NEXT FORCE IS cognition. We are going to define it in the broadest sense we can as the act of using information. Our mammoth hunters employed their cognitive abilities in several ways all throughout that story.

There are so many terms surrounding cognition that it is easy for them to all muddle together. After all, what's the difference between thought, calculation, intelligence, reasoning, mental processes, problem-solving, and a litany of other terms? It's really a mess, but the good news is that it's not a mess we have to clean up. It's irrelevant to our inquiry. Why is this?

Our goal is to understand the forces that combine in superorganisms to produce emergent capabilities. We want to go broad in our understanding of cognition so that we can find the commonalities among cells, bees, beehives, people, and Agora. It turns out that what they all have in common is their ability to put information to use.

We've explored how organisms turn data into information, as well as the many ways that they can store, access, and share it with others. What we skipped over was how that information is then used. That's cognition.

Cognition involves the selection of which information to use, the actions to perform based on that information, and the modification of that information based on new data. There's a specific kind of cognition that we are really good at called reasoning in which different pieces of information are combined to create new information.

Because we are dealing with such a root-level ability as cognition—as opposed to something much more complex like creativity—we can't really turn up our nose at a cell's cognitive abilities. It is, after all, where your hundreds of megabytes of DNA reside and are acted upon. We would never say a cell reasons or is creative, but we have to give credit where it is due: Cells process a great deal of information. While they have few ways to acquire new information, they are the custodians of the billions of years of accumulated information in DNA, and they constantly perform actions based on that information.

I belabor the point because we are going to see a range of amazing abilities from creatures that are just acting on instinct, performing actions coded in their DNA. But we can't say those are necessarily any less cognitively rich than actions performed by reasoning creatures such as ourselves. Again, we are only concerned here with a single dimension: cognition, which we are defining as using information. Even nonliving systems can perform cognition in this sense by simply being able to adaptively act on information. Does that mean they can think? Or they are intelligent? I don't think so, but that's an open debate that needn't distract us. Can computers be intelligent? Or can they merely emulate intelligence? Or is there even a difference between those two things? That's a rabbit hole we'll explore in another book.

We can't really say which of our six forces is the most important because they are all necessary, except technology. That would be like asking which is the most important wing on the airplane, the left or right. That said, cognition is the one that pushes us upward in complexity, creating new levels of order. It's the transforming one. You could read and write all the information you want, communicate night and day, specialize to any degree, and expend energy, but unless you are doing something with the information, then yesterday, today, and tomorrow are exactly alike.

I don't use this broad definition of cognition to dilute the idea into something meaningless. Not at all. The idea is that if something can act on information, it doesn't matter whether it is smart, aware of what it is doing, alive, or conscious, or anything else. What we see then is that cognition is everywhere.

And given that it is the method by which energy is deployed to create order, then order is everywhere, too. That explains the vast number of superimposed systems—both natural and humanmade—that exist in the world. Recall that systems definitionally have feedback loops, that is, ways to acquire and use information. Therefore, all systems are cognitive systems. Others may define all these terms a bit differently than I do here, but we can all agree that systems use information, and that doing so creates order.

So, with that preamble, let's look at a few cool examples of cognition in the hopes that we will spot similar forms of these at play in Agora.

- Termites eat non-load-bearing wood before they eat the load-bearing parts so that whatever they are all munching on doesn't come crashing down on them.
- Your immune system recognizes many different pathogens and attacks them, leaving friendly cells unharmed.
- When its leaves are being eaten by caterpillars, a certain wild tobacco plant releases a scent that attracts the caterpillar's predator.
- There are hundreds of species of spiders that pretend to be harmless ants instead of bloodsucking monsters by holding their two front legs up in the air a bit to make them resemble antennae, leaving them with just an ant's six legs.
- My personal favorite is the tale of the humble philodendron. Its life begins as a seed in a bird dropping deposited on the ground in a rainforest. It sprouts and immediately starts heading toward the nearest tree, which perhaps it can identify because of some quality of light in that direction. By "heading to" I mean that the part of the plant closest to the tree grows toward it while the part farthest away withers. Sped up, it would be like a six-foot-long snake slithering toward the tree, then climbing it. When the plant reaches the canopy, it tries to travel to a different tree by throwing out a shoot. If it misses and falls to the earth, well, it starts over. And all along the way, it grows different-sized leaves to meet its energy needs as it goes through the various stages of its journey.

- Finally, some animals perform what are called fixed-action patterns. Consider the graylag goose. If one of its eggs rolls out of its ground nest, it will use its bill, neck, and chest to maneuver it back with the other eggs. If, while the goose is doing that, you reach down and take the egg, the goose will still keep going, trying to roll an invisible egg back to its nest. This is an interesting one because the goose is reacting to one kind of information—the egg rolling out—but oblivious to the other, that the egg has vanished. Evolution never saw a need to cover that latter scenario either because it seldom happens, or, more likely, it doesn't further survival. Putting the egg back does further survival, but knowing when the exercise is pointless doesn't.

There's a big question we haven't addressed yet, which is how creatures know what information to act on, and, for that matter, how to act on it. What information should they combine with other information? And most perplexing of all, how do creatures with minuscule cognitive abilities know how to act on relatively complex information?

To answer all this, we need to understand algorithms. An algorithm is a series of steps designed to achieve some goal. A recipe is an algorithm to make a cake. Algorithms don't have to be linear—that is step-by-step—but instead they can be conditional, like a logic tree. An algorithm to tell a cold from the flu might be: *Do you feel sick? Yes or no? If yes, do you have a fever? If yes, then you have the flu, otherwise just a cold.* OK, that's not a very good algorithm, but I trust you get the general idea.

Earlier we mentioned a kind of cognition known as reasoning in which information is combined to create new information. Algorithms are a form of that. But only a few things can reason, so everything else must be born with their algorithms, and new algorithms must be developed via natural selection, a tedious process that can take millions of years. The big idea here is that by combining simple algorithms you can get incredibly complex results. One example from the bee world illustrates this:

As mentioned earlier, bees cannot regulate their body temperature and are thus termed cold-blooded. However, the superorganism that is the bee colony does regulate its temperature, so it is a warm-blooded creature. The colony's ideal temperature is in a narrow range, quite close to our own body temperature in fact. How do the bees use simple algorithms to regulate the temperature of the colony? It turns out that bees have one algorithm for cooling the hive and another for warming it. To cool it, they go fetch water, bring it back, and spread it on the comb. The water then evaporates, cooling the hive. Other bees' algorithms place them at the entrance of the hive, flapping their wings to pull out the hot humid air. When the hive is too cold, the bees have a way to flex their muscles to generate heat. It is calorie intensive, but it works.

The way that bees regulate the hive's temperature using these mechanisms is fascinating. It turns out that every bee has a slightly different temperature preference, just like humans. If the hive gets just a little warm, a couple of bees are all like, "Man, it's hot in here. I'm going to cool this place down," and they fly off to get some water to spread on the comb, cooling things down a bit. The other bees look at each other like, "I'm not hot," and they do nothing. But say the temperature keeps rising. Then more bees say, "You know, it is getting stuffy in here," and so they start fetching water. Eventually the temperature falls a little, causing those latecomers to stop working, but the original ones still find it hot, so they keep at it until the temperature gets to a point that everyone is happy with. So while the climate control system in your home is binary—either on or off—the hive's system is analog, gradually kicking it, then gradually turning off. All of this happens with no planning, in a bottom-up fashion driven by the simplest algorithm. Any bee has just three states: hot, cold, and just right, and since all the bees vary slightly, you end up with a very nuanced, complex ability.

We see these sorts of algorithms all over. Toads, for instance, will regard a moving horizontal line as prey and will try to eat it, even if it doesn't look like food. However, take that same line and orient it vertically, then parade it in front of the toad, and he won't give it a second look. Algorithms like these don't have to be perfect, but they can often provide an advantage by simply being right more often than wrong. While the simple algorithms are quite robust, when they fail, it is often catastrophic. Army ants can mistake an old chemical trail for a current one and accidentally start walking in a giant circle. The more they do that, the more it reinforces the trail, causing them to double down on their commitment until, in the end, they walk themselves to death. Likewise, the Australian jewel beetle's population became threatened because whatever simple algorithm it uses to identify a potential mate caused males to identify a certain brand's beer bottle as a particularly attractive female, and they tried to mate with it.

Of course, it isn't that some things are born with algorithms and some learn them all. It's a continuum, with us on the learning end and cells, for instance, on the "born with" side. In fact, almost everything is way over on the "born with" side. A human baby is born with just a handful of algorithms, such as how to spot their mother's breast, and must learn everything else, a process that literally takes a lifetime. On the other hand, snakes are born knowing almost all they need to know in life. As are baboons, badgers, and bonobos. Animals learn little in their lives because of their relatively limited ability to accumulate information and create algorithms to navigate it.

So that's our fourth force, cognition. All superorganisms must have a form of cognition to make use of its information stores, and evolution has created a litany of ways to do this. Cognition can occur within the individual parts or within the superorganism itself. For instance, with humans, your parts—the cells—are largely self-sufficient, dealing with their problems on their own. But you, the emergent superorganism, also performs cognition, and, as we will see, so does Agora.

SPECIALIZATION

OUR NEXT FORCE IS specialization, also known as the division of labor. We can see our mammoth hunters employing it to great effect, and in fact all superorganisms utilize it. If all of the parts are just clones of one another, all performing the same tasks, then a group of them is just bigger, but it is no "better." That is, it isn't going to achieve any emergent properties. We saw that in our tug-of-war example.

The division of labor is a simple but amazingly powerful idea: If we all specialize, we can all be better off. The idea is probably best known through Adam Smith's famous pin factory example from the 1776 book *The Wealth of Nations*, where he said that a certain pin-making factory that employed ten men could make one hundred pins a day if the men each worked alone doing all parts of the job, but if they specialized and each did just one part of the task, then the same ten men could make 48,000 pins, a nearly 500-fold gain. But this idea predates Smith. There's a long discourse about it in Plato's *Republic*, where Socrates argues that "all things are produced more plentifully and easily and of a better quality when one man does one thing which is natural to him and does it at the right time, and leaves other things."

But doesn't this seem like an improbable amount of productivity gain? Smith says there were just eighteen different steps in making a pin. Instead of everyone doing all eighteen, in the new arrangement, each person did just one or two. Wouldn't it seem, at a gut level, that maybe they would see a tenfold increase in efficiency? Or maybe a twentyfold? But how do they get a nearly five-hundred-fold increase? There are a few things going on here. The first is the obvious one: proficiency gains. A person who tries to master one sport, say golf, will be vastly better than someone who tries to master eighteen. In addition, though, the pin factory will hire specialized people to do the specialized jobs. One job might require someone with fine motor skills, another might require a stronger person to cut and manipulate the wire, and a third job might be done better by someone nearsighted to see the point on the pin. Consider your various organs. You have a heart, lungs, liver, and stomach, all of which are highly specialized. Imagine if evolution had to make a single organ that did what those four organs do. It wouldn't be just a quarter as efficient as the specialized ones but only . . . well, who knows, actually? It might not even be possible, and I can guarantee it wouldn't be pretty.

Smith may have taken a bit of poetic license with his pin example to make a—ahem—point, but in principle I think he's right on.

Carl Sagan once said, "We live in a society exquisitely dependent on science and technology, in which hardly anyone knows anything about science and technology. This is a clear prescription for disaster." As much as I admire Sagan, I have to part company with him on this one. I think it is pretty cool that someone can invent a cell phone and that I can use it to talk to anyone on the planet without understanding anything at all about how it works. Imagine if you had to repair your own car, make your own paper, and grow your own food. Almost every object in my life is of higher quality than I could have produced myself. Just last week, I tried to decorate a cake for one of my kids' birthdays using one of those squeezy-bag things to apply the icing. I mean, look at me; I don't even know what they are called. As you may have guessed, my decorated cake didn't turn out well, but my wife, surveying the wreckage in the kitchen, said I had done a pretty good job frosting everything else. I'll stick to writing and leave the cake decorating to the cake decorators.

You probably know the phrase "a jack of all trades," but often the back half of that is left off: "but master of none." The division of labor works because we individually become masters of some trade and don't even attempt the other ones at all. The curious thing about the division of labor is that when we specialize, we individually know much less. I do only a few things well, but there are many thousands of things I do quite poorly, even worse than cake decorating. However, even though the average knowledge might fall when people specialize, the collective knowledge grows, because it is the aggregate of all the different things that everyone knows.

The benefits of specialization don't even require us to be in proximity with one another or even know about one another. This was wonderfully illustrated in a 1958 essay called "I, Pencil" by economist Leonard Read. In it, Read points out that no one on the planet knows how to make a pencil. Who among us can mine iron ore, smelt it, turn it into steel, and form the ferrule? Even if such a person were alive, they would then have to find a rubber tree, learn how to tap it, and—well, you get the idea.

But wait. Didn't someone invent the pencil we have today? Surely, they must have known how to make one, right? No, not at all. That inventor just snapped together parts other people made, who in turn got their supplies from many other people, who had suppliers who got their raw materials from yet other sources.

This does raise the question of, "Well, if no one knows how to make the pencil, how do they get made?" Answer: Agora makes them. A smartphone, which contains roughly twice as many different elements from the periodic table as your body does, is exponentially more complex than a pencil. Not only can no person *make* a smartphone, but I would wager there is no one alive who even *knows* how a smartphone is made. There is no one who understands the entire process down to identifying a useful vein of ore in a mine. Our society is so specialized that there are probably thousands of different jobs that each require an entire career to gain enough mastery to make that phone, all the way down to the programmers, accountants, lawyers, and all the rest. It is literally impossible for anyone to know more than a minuscule fraction of how to make anything that is the least bit complex, let alone all of it. So, Agora is making our smartphones as well. We'll dive deeper into that in section III.

All superorganisms are specialized and, in general, the number of different roles in a superorganism is proportionate to its complexity. Bees and ants have only a handful of different jobs, yet they achieve amazing emergent properties. Even you, an incredible superorganism, are made of just two hundred different types of specialized cells. But as we will see, Agora is composed of tens of thousands of different specialized parts. This extreme specialization is not taxing on the superorganism for the simple reason that no part of the superorganism is keeping track of it all. Not only is there no bee that knows how to do everything to run a hive, the bees don't even have any idea what's going on elsewhere. It is all decentralized.

Being decentralized, the parts need to be largely autonomous. Johann Wolfgang von Goethe wrote, "Let everyone sweep in front of his own door, and the whole world will be clean," and in a superorganism, all the parts just sweep their own front stoops and mind their own business. In our beehive, each bee just does its job without any direction. The bees have no manager, no one who watches to make sure they neither dawdle nor dillydally. The queen bee is not a ruler—that's just a metaphor. Most of the bees will never even meet her. Likewise, no one in the mine hangs around waiting for a call from Tim Cook telling them where to dig next. Just imagine if you had to direct the activities of each of your trillions of cells. You'd be up all night. This autonomy adds resilience to the organism.

None of the parts of a superorganism specialize for altruistic reasons. They aren't driven by a desire to "do their fair share" or "be a good team member." Remember, they don't even know that the superorganism exists. A part specializes to further its own survival. No one decides to become a better accountant so Agora can be better off, but so *they* can be better off.

Picture it this way: Pretend there are two selfish bees, each of whom only looks out for number one. If, while living their selfish lives, they just happen to benefit each other, then a partnership is formed. But even then, the bees continue to look out for just themselves, indifferent and oblivious to the effect they are having on each other. And so maybe the partnership dissolves at some point. But more likely, each bee continues to optimize its own life by doing more of whatever was working, benefitting the other even more, but again, oblivious to that fact. Eventually, the advantages of the partnership are so overwhelming that there is no scenario where the bees are better off outside of it.

That's coevolution, and that's the beginning of the superorganism. It's also the mechanism by which specialization happens, and the feedback loop described earlier keeps pushing toward more and more of it.

Thus the collective becomes more specialized without any intent from any of the parts. It's not a "strategy" of theirs, although colloquially it might be described that way. Impersonal forces incrementally drive it and shape the evolution of the individual bees. But this specialization ultimately has a trade-off. As they specialize, they are less able to survive on their own. They have thrown their lot in with the superorganism and have been better off every step of the way. So it's a good deal, really. This applies to everything from the cells in your body—none of which can survive on their own—to the social bees that quickly die if they happen to be out running errands when the beekeeper relocates their hive, to, as we will see later, individual humans.

TECHNOLOGY

THE FINAL FORCE IS technology. It is the only one of the six that is optional. Relatively few superorganisms have it. Its power is that it is a force multiplier; it amplifies everything else. With it, we can get more energy, acquire more information, communicate better, augment our intelligence, and specialize better. We increase our own productivity with it, and so if Agora exists, then it amplifies Agora as well, via us. Our mammoth hunters had a range of it.

For practical purposes, humans got technology when we got speech. While we had rudimentary tools before then, the archeological record unambiguously shows an explosion of technology at the same time we began making representative art, an ability we suspect came with the development of language. These early tools demonstrated a shocking amount of sophistication, including construction from multiple materials sourced from diverse locations. The tools, just like us, also became more specialized over time. Gone were the good old days when you could get through life with just a pointy rock in your pocket.

Recall the cave paintings and how they seemingly came out of nowhere—that is, had no precursors. Not only do these paintings exhibit artistic excellence, but they were executed with astoundingly sophisticated technology. In Lascaux cave in France, we find beautiful paintings that date from almost twenty thousand years ago. The paintings clearly must have been executed with scaffolding given how far off the ground they are. Because the paintings are so deep in the caves—sometimes a full mile in—there also had to have been some kind of artificial light source. Plus, the artists mixed their pigments with animal fat to get them to adhere better and with talc to extend them. As impressive as all that is, wait until you hear how the artists got black pigment. Understand, they had plenty of things that were "blackish." They obviously had charcoal and even soot in abundance, but both of those are more gray than black, and evidently the artists insisted on having a true black.

To get that, they needed an uncommon mineral called hausmannite. This is notable because to turn it into a pigment that can make true black, you have to heat it to 1,600 degrees Fahrenheit, which is not easy to do in a campfire—that's 400 degrees hotter than the melting point of aluminum. And, get this, the closest known source of hausmannite is over 150 miles from the Lascaux.

Hold that thought for a minute.

Earlier we discussed the half-century-old Moore's Law, the observation that the power of computers doubles every two years. When Gordon Moore first noticed it, he believed that it had been going on for a few years and would probably continue a few more. But computers, or more technically, calculating devices, have been around since the late 1800s. What Gordon Moore didn't realize—and there is no reason he should have—was that this doubling had been going on since those earliest machines. What makes that so astonishing is that computers themselves completely changed multiple times throughout that period. Their core technology went from electromechanical to relays, then to vacuum tubes, then to transistors, then to integrated circuits, and finally today's microprocessors. Those technologies don't have anything to do with one another, and yet Moore's Law never even hiccupped. How can that be? How can an abstraction behave in such a law-like manner?

The mystery deepens when you learn that many other technologies follow Moore's Law as well. They don't necessarily double every *two* years, instead they double every *x* years, that is, on a consistent, periodic basis. Futurist Ray Kurzweil, to whom we owe this brilliant insight, has identified a range of technologies that behave this way, including cases relating to data transfer rates, computer storage, megapixels in cameras, all the way down to nanotechnology. The list of places where we have seen this doubling occur grows steadily year by year. The complexity of multicellular life, as noted earlier, seems to double about every 350 million years.

Not only are our technologies themselves improving, but our technical knowledge is accumulating as well. This is why so many technologies are invented concurrently. The light bulb was 99 percent invented before Edison was born. We already knew how to make glass and refine metal. We had laws for patents, a financial system that allowed for raising capital, and thousands of years of accumulated scientific knowledge. All Thomas Edison or Alexander Graham Bell or Henry Ford did was just the last little bit, and then they waltzed in and claimed credit for inventing the ice cream sundae when all they did was put the cherry on top. I'm not trying to diminish their accomplishments, just set them in context, just as Isaac Newton did when he credited his own achievements to the fact that he was only able to see further because he "stood on the shoulders of giants."

I have avoided defining technology up to this point, trusting we all know more or less what we are talking about. But now, we need something just a bit more precise. While we tend to think of technology as gadgets and tools, that isn't quite right. The essence of technology is knowledge—specifically, applied knowledge. *Knowing* how to make a stone ax is as much technology as the ax itself. So as our knowledge accumulates, our technology does so as well, almost by definition. Thus technology is a bit like compound interest; as you have more of it, you can use it to make even more, growing it ever faster. This is what powers Moore's Law. Or put another way, at 7 percent interest, money will double every ten years. Even after several doublings, it still only takes ten years to double. But those later doubles brought in much more money than the early doubles. Technology behaves this way, and ours has doubled so many times that each subsequent advance is a vast amount of increase.

The concept of technology is a big tent, encompassing many diverse things. Your grandma's recipes are a form of technology, as are fables, law codes, our knowledge of the curative powers of herbs, our customs that allow us to get along better. It includes all products of the human mind where we take something we have learned, that is a piece of useful knowledge, and embody it in some way, either as a device, a paragraph, a story, a procedure, or a rule of thumb. While this may seem like a dilution of the concept of technology, broadening it so much that it loses all useful meaning, the opposite is actually true: the takeaway is that technology is so pervasive that it is like the air we breathe.

If Moore's Law is as ubiquitous as I suggest, shouldn't all that be getting constantly better, too? Doubling on a periodic basis? Laws, fables, and all the rest? Now that is the question, isn't it? The short answer is yes, but we'll deal with that question extensively in the next section when we try to understand why progress happens.

We explored earlier how humans have to be taught everything we know; that we aren't born with it. This is a process that starts when we are young, as we learn how to wait our turn at the water fountain, write our name, and brush our teeth, all three of which are technologies. As we age, we learn a hundred thousand more pieces of technology. How to form an argument, balance a checkbook, tell time, walk on ice, rake leaves, do the dishes, write a memo, and so many more things of incredible subtlety, like determining whether to use an exclamation point in a text message. Our society requires so much knowledge to maintain that it takes years and years of schooling for a new human to become a contributing member, and even after all that effort, they have to keep learning new things throughout the rest of their lives.

Luckily, we are suited for learning new things, being vastly better at it than computers. While we have tried to train a computer to drive a car by using millions of miles of data and billions of dollars of expenditures, success has eluded us. Yet a human teenager can learn to do a decent job in about twenty hours.

Because animals learn so comparatively little throughout their lives, they have relatively little knowledge to accumulate and pass down from generation to generation, in the way that we pass down the knowledge of how to tie a square knot or make chicken and dumplings. That's why one generation of beavers is no different from the previous hundred nor the next hundred, for that matter.

A beaver can construct a dam and a bird can build a nest, but these aren't actually technology. They are just biological processes, like digestion. If you look at digestion as a complex system, you might be inclined to say, "Man, that beaver is brilliant. Look at the amazing way he digests his food." But of course, we don't say that. Digestion may be of extreme complexity, but we don't credit the beaver's intellect with doing it. The beaver building a dam is doing the same sort of thing, just externally. The beaver doesn't know what it is doing. If on one fine spring morning, a beaver is strolling through a field and comes across a device playing the sound of running water, the beaver will drop whatever it is doing and build a dam over the device until it no longer hears the sound. So, we can't really say the beaver uses technology when it builds a dam even though a human building the same exact dam would be employing technology.

Wait a minute, though. Maybe this seems like a double standard. How can we say that a beaver, which can build a dam so well that it will likely change the local ecology, *isn't* using technology, but that the three-year-old learning to brush their teeth is?

This is a valid question. After all, they both look like procedures, and procedures are clearly technology, one of the most common kinds. But if technology is the practical application of *knowledge*, then the creature—human or animal—must *know* what they are doing, which the beaver doesn't. This isn't me just splitting linguistic hairs. If evolution, over countless eons, just dumbly eventually stumbles upon a procedure that works, well, that's not technology. The child on the other hand is taught knowledge. Brush lightly back and forth, at the gum line, with the brush angled this way. If beavers *learned* how to build dams, then sure, that's technology. But if it's instinctual, then we are back to something more like digestion.

I don't mean to say that animals never learn new processes or techniques. They do it all the time. The internet is full of cute videos of the old dog showing the new puppy how to go in and out of the dog door or go down a flight of stairs. I'm a sucker for them, frankly. I've seen ones with a mama cat using her paw to show a kitten how to the drink from a water dish, as well as videos of both cats and dogs going through long procedures to get a cabinet door open or reach some high spot. And then, of course, there are service animals that learn incredibly complex procedures. Operating a dog door isn't something that a dog instinctually knows how to do, so the technique they learn is technology. Rudimentary, yes; but it absolutely is the same sort of thing as knowing how to build an atom bomb.

Eventually, in one hundred thousand generations, these animals may evolve much more sophisticated technology, but for now that's where they are. And this explains a question that I have gnawed on for years: How can a beaver build such a magnificently constructed dam, something that sure looks intelligent, and yet doesn't know how to do anything else with those exact same skills? It doesn't have general knowledge on how to build something, just an instinctual drive to build a dam.

But let us not be too smug here. Humans also do things just like the beaver building a dam over the recording of water. The way our brains operate, though, we would rationalize a reason for doing it. "I don't like the sound of water," we would tell others if they questioned our motives. We've already seen this in action: Recall the passport photo story from earlier. The one where people made up reasons for selecting a photo that they didn't really select.

Is it possible to live without technology? Not really. Groups of people that explicitly reject technology for a simpler lifestyle are really only rejecting one narrow sort of technology. None of these groups, for instance, disavow the technology of clothing or metal, let alone speech. They don't eschew the technologies that allow them to build shelters that don't collapse or grow their own food. Unless they are basically feral humans, walking around naked and flipping over logs looking for grubworms to eat, they are employing technology in their lives.

Thus technology is a form of order that we create. It's different than other forms of order in that it accumulates, not dissipates. Since it accumulates over time, we have technologies we have been fiddling with for millennia, and each of those might have millions upon millions of aggregate person-years of development in them. How can that be, though, if behaviorally modern humans have been around only a fraction of that time?

When you consider the technology in a car, that story doesn't begin with Karl Benz's 1886 patent. It doesn't even begin with the Bronze Age. It really goes all the way back to the beginning of perhaps our greatest technology, speech. If that was fifty thousand years ago, then wouldn't that be the amount of technology in the car? Fifty thousand person-years? No. The technology to build a car took a thousand separate paths to come into being, and they merely converged in 1886 to make that car. One path brought agriculture that fed the people who made the bronze. Another path lead to the development of the monetary system, another led to the legal system, and so on. And if you take those thousand paths, each of which took thousands of years, then in aggregate Benz's first car already had countless millions of years of technology in it. That ain't pushing a doggy door open with your nose.

As we embodied technology into objects, from hammers to Humvees, we expanded our own capabilities. We became supermen and superwomen. With technology, we became faster than a speeding bullet; the SR-71 Blackbird flew at 2,100 mph while a bullet lollygagged along at 1,800. We became more powerful than a locomotive, and while jet packs certainly aren't mainstream, a few of us can in fact leap the tallest buildings in a single bound.

Recently there has been a trend toward making technology more autonomous so that humans don't inadvertently become the bottleneck in our own progress. We enhance its autonomy by giving it sensors, that is, means to measure specific elements in its environment. Additionally, we build this technology in such a way that it can communicate, not just with us, but also with other pieces of technology. This is new. Hammers don't talk to nails, nor do they converse with saws. But that is all changing.

And yet, we still have a long way to go. Our world remains pretty low tech. Why do I say this? Many of the tasks that humans do require us to imitate machines; that is, do work that machines could in theory do. We may use human labor in lieu of the machine because we don't have the machine, or labor is cheaper, or the machine hasn't even been invented yet. This is really a waste of human potential, because when you make a person do a task that a machine can do, then you are basically dehumanizing the person, that is, regarding them as a piece of technology—a mere machine. You are saying, "I don't need any of your unique human characteristics. I just need you to act like a machine I do not own." Any work that you can picture a machine doing autonomously is work that is beneath the dignity of a human, the most exquisite, extraordinary, amazing thing in the known universe.

Eventually, we will have machines to do all the machine jobs. This will unlock human potential and trigger a great flowering of civilization. Right now, someone may be an amazing poet, but we need a hole dug, so we tell that person to stop reciting and start digging.

I throw my lot in with Buckminster Fuller, who said, "The true business of people should be to go back to school and think about whatever it was they were thinking about before somebody came along and told them they had to earn a living."

Eventually, we will get to the wonderful world that author Warren Bennis had in mind when he wrote, "The factory of the future will have only two employees, a man and a dog. The man will be there to feed the dog. The dog will be there to keep the man from touching the equipment."

CONCLUSION OF SECTION II

SO THAT'S HOW THE magic happens. Superorganisms are powered by energy and are able to gather and store information. Their parts interact with that information via cognition, usually in the form of simple, instinctual algorithms. Through specialization, natural selection is able to optimize multiple attributes concurrently. It doesn't have to build a single uber-ant that can do everything; in fact, it can't. Try as it might, it can't build an ant that is both large and small. Instead, it builds different sorts of ants, each manifesting the forces differently, powered by their own set of specialized algorithms. When those ants interact, they form the emergent uber-ant, but in a distributed form. We call those hives. Likewise, instead of trying to build an uber-hive, natural selection may well be making different sorts of hives, each with different attributes that collectively will form the distributed uber-hive. All of these various levels of life, right down to the individual cells of individual ants, are superimposed upon each other, and are all evolving in parallel, in a symphony of life.

But how does this explain the termites eating load-bearing wood last? I hope that by considering the interaction of our six forces (or five, actually, given that they don't have technology), you can see how they could, *in theory*, bring about that outcome. "But come on," you might be thinking, "how could that specific capability have mattered enough times in the distant past to have altered the evolution of an entire species?" That's the final trick.

The easy answer, and it is undoubtedly true, is that humans lack frames of reference that allow us to comprehend the vast expanses of time over which this all occurred, as well as the unfathomable numbers of termites that have lived and died. There may be one hundred trillion termites alive right now, each of which might live a year, and they have been around one hundred million years. There was no reason for evolution to give us the ability to understand either of those numbers, let alone the result of multiplying them together.

But there is another, more subtle reason that such an improbable capability came about: We also cannot perceive the incalculable number of advantageous behaviors that never evolved or at least never persisted. In other words, there are all sorts of tricks termites might have learned to help their species that just never happened, so it is less of a wonder that one particular one, however unlikely it may be, happened to be stumbled upon. Think of it this way: Imagine you have a hat that has slips of paper numbered from 1 to 1,000 in it and you pull out a random one that has 546 on it. The odds of you pulling out number 546 were 1 in 1,000. Highly unlikely! But that would be true of any of the numbers you pulled out, such as 65 or 912. They, too, are highly unlikely. Knowing to eat load-bearing wood last is just #546. It's the one that just happened to get drawn from the hat.

So here we are: In section I we saw the development of life and its journey from relative simplicity to unfathomable complexity, all the way up to planetary superorganisms. In section II we went the other way, all the way down to the fundamental forces that come together to create superorganisms. We are now completely equipped to answer the question we all came here for: Do humans form a superorganism?

That brings us to section III—Agora. We'll begin by describing Agora in more detail and seeing how our six forces come together in the parts—that is, us—and produce such an emergent entity. We'll see all the ways that we resemble the superorganisms we have looked at in such detail. We'll see how we spend our lives buzzing around our hives—our cities—driven by our simple algorithms, forced to conform to the standards of the hive, and unable to survive apart from it.

We'll come out of that tackling some of the questions that we've mentioned along the way. What does the superorganism want? What is its purpose? Will it grow old and die? Can it reproduce? And finally, why are we here?

In closing, I'm reminded of an episode of *Star Trek: The Next Generation* in which the leader of an alien planet learns of the existence of the larger universe—and of the countless other species it contains. He tells Captain Picard that every evening when he gets home from work, his wife and two daughters ask him over dinner if he had a good day.

> **Captain Picard:** "And how will you answer them tonight?"
> **Chancellor Durken:** "I will have to say: 'This morning, I was the leader of the universe as I knew it. This afternoon, I'm only a voice in a chorus. But I think it was a good day.'"

If after finishing this book you end up coming to believe that Agora is a real, living creature, and that you are but a small part of it, how would that make you feel? Would it change your view of the world and your place in it? And if so, would you think it was a good day?

SECTION III
AGORA

MEET AGORA

AGORA IS A SUPERORGANISM composed of each and every human. Why did I name it Agora? In ancient Greece, the agora was a large, open, outdoor area that was the heart of the activity of a city. Picture a noisy, crowded marketplace where sellers hawk their wares to passersby. Then superimpose on that image a debating society working through the issues of the day. Then imagine that the courts are there as well, along with government representatives and religious leaders. Picture a town crier or two walking around shouting some news of some kind, trying to compete for attention with the buskers playing their music. Imagine the aromas of all the food being cooked by the vendors, superimposed on the odors of the animals brought to sell, combined with the musky scent of a thousand humans who only bathe on their birthday and who may have missed the last one.

Herodotus, the ancient Greek historian, writes about a custom that he admired of the Assyrians in their agoras: "Having no use for physicians, they carry the sick into the marketplace; then those who have been afflicted themselves by the same ill as the sick man's, or seen others in like case, come near and advise him about his disease and comfort him, telling him by what means they have themselves recovered of it or seen others so recover. None may pass by the sick man without speaking and asking what is his sickness." It's like posting your symptoms to an online forum and having people weigh in on what you should do. So now imagine the sick laid all about, an auction or two going on, a couple of thespians acting out a scene of an upcoming play, and a heated argument over the relative merits of single-humped camels versus double-humped ones, a topic that even Aristotle had an opinion on. If you listened carefully, you would hear multiple languages spoken, and if you looked around, you would see people in all manner of dress from distant lands in every direction, all of whom brought goods to trade.

And it is that image of the agora, with all of its interactions between people, that prompted me to choose it for the name of the superorganism of us.

Agora is a planetary superorganism, like Gaia, which we discussed earlier. It is a living being whose cells are people and whose hives are cities. There is nothing particularly mystical or supernatural about Agora, or, more precisely, there's nothing any more mystical and supernatural about it than the phenomena of life and emergence. Agora is no god, but a mortal creature, just one that operates on a different scale than we do—in both time and space—and it exists in such a way that we cannot directly perceive it, the way our cells cannot perceive us.

Agora was born with the city, emerging from the activity within it. The earliest cities were Agora in its simplest form. As those cities grew in size and complexity, so did Agora, and the civilization that we have so painstakingly built over the years is Agora's maturation.

Agora includes all the world's people who are connected to one another, which today is virtually everyone. The entire world now functions as a single metroplex, and so Agora has grown to be a planet-scale superorganism. Not only is it *all* humans, it is *only* humans. It isn't a superorganism of all life—that would be Gaia—but is a purely human creature of which plants and animals are not a part, nor is technology.

Just as our minds are comprised of many processes—you will recall our brain's "drinking a cup of tea" example from the chapter on the mind—Agora's is as well. It has a whole lot of stuff going on all at once in its metaphoric head. There are three brothers in Brazil trying to get a truck out of the mud, fourteen people in a conference room in Montenegro trying to come up with a name for a new mouthwash, and twelve thousand people at a stadium in Mexico City doing the wave. Each of those, and a hundred million more, is akin to a human mental process, and collectively they form Agora's mind. They may seem unrelated, but they aren't, because they are all part of the same creature. To understand that in human terms, imagine you are working a newspaper crossword puzzle and give yourself a paper cut. While your mind is still working on the puzzle, your platelets are racing off to stop the bleeding. Those are completely unrelated things except that you are the one doing both of them.

How does Agora manifest? In other words, can you see Agora? Or feel it? That's an interesting question. Consider that first empty pine bee box that I ordered from the Sears catalog all those years ago. Was that a superorganism? No, that was just a wooden box. Is that shipment of bees that was delivered to me a couple of weeks later a superorganism? No, that was just a bunch of bees. After the bees were all placed in the wooden box, was all that together a superorganism? Yes, that's much closer. As the bees work in the wooden box, they transform it, the same way Gaia theory posits that life transformed our dead planet. The bees build passageways and honeycomb walls, and they designate places for different tasks. This is a form of encoding information, and every time one of those bees walks through a passageway, it is a form of cognition—a memory being accessed. And it is that whole system that the bees work so hard to keep at homeostasis that is a superorganism.

When I say that technology is not part of Agora, that's both true and not quite true. Technology is like that empty bee box; it is instantiated knowledge. We built the bee box a certain way for a certain set of reasons, accumulated over the eons that humans have kept bees. The mosaics of Pompeii depict bee boxes hardly different from our own. The bee box is a manifestation of human knowledge, a.k.a. technology. Cities are Agora's hives, so an empty city can be thought of as an empty bee box. We can casually refer to it as part of Agora, and it is, but it is important that we not confuse that with the human activity that creates Agora. We are the only living components of Agora; we are the heart and soul, but the world's infrastructure upon which we live can collectively be thought of as part of Agora as well. Maybe this seems like an arbitrary distinction, but think of it this way. Consider a person with an artificial arm. Is the arm part of the person? Well, yes and no—it really depends on context.

Because it is under the person's control, it can sometimes be useful to think of it as part of them, but it really isn't, any more than a pair of glasses is. That is roughly how Agora relates to technology.

What powers Agora? We went into this in depth in the energy chapter, but a quick refresher: Superorganisms don't have to have any power source beyond their parts. After all, a beehive runs on bee power, which is really just cell power. Agora is a bit different. The power at its disposal is the collective energy of eight billion human bodies, but also all the energy generated by humans with our power plants and refineries and the rest. That first number, the energy that we collectively get from food, is about twenty-five exajoules of energy a year. However, humans generate and have at their disposal an additional five hundred exajoules of energy per year, which means that Agora is powered by 525 exajoules a year. As a reference point, the amount of solar energy that the sun rains down on this planet is about four million exajoules, so on a cosmic scale, Agora is a small creature.

Agora is a living, complex creature, and is structured like other complex life—including humans—as a set of superimposed systems. We can even map Agora's systems to the sorts of biological systems we are all familiar with. However, we must be careful here. This mapping is little more than a metaphor. Superorganisms never resemble their parts. A beehive doesn't have a stinger, and you look nothing like your cells. That being said, I think there are natural parallels that have helped me visualize how Agora operates, that is, how the actions of the parts—us—work together to create the whole.

Musculoskeletal system: Our cities, with their hard infrastructure such as roads and buildings, function as Agora's skeletal system. That's the pine bee box. Agora's muscles are everything that we make that moves stuff, including all the dozers and cranes, along with the ships, cars, trucks, and trains.

Endocrine system: Hormones are sent around the body by cells to alter the behavior of other parts. Money is an example of such a hormone. If I pay you a dollar to do something, then I altered your behavior with that dollar.

You might recall from earlier how ants use pheromones to signal other ants to change jobs when there is a shortage of one sort of worker. The human equivalent works like this. Someone runs an ad looking for a waiter and offers a wage. If there are no takers, they send out more of the money hormone—that is, they offer a higher wage—until someone leaves their job to become a waiter.

Digestive and excretory system: The global food supply chain is the digestive system of the superorganism. Our sewage system is the excretory system.

Circulatory and cardiovascular systems: Agora's bloodstream is the world economy and all the various supply chains. It is everything that gets made and moved around. It requires enormous amounts of raw materials to keep it running, everything that we grow or mine. It also generates vast amounts of waste, which includes our landfills and our environment, that is, everything we emit into the air and water.

Now that we have met Agora, let's try to get to know it.

HUMAN HIVES

EARLIER, WE EXPLORED A theory on how bees banded together to become a superorganism. Over time, it is thought, as they lived in ever-closer proximity to one another, some amount of specialization occurred, which gave them an evolutionary advantage over other bees. At some point, the heap of bees began to display emergent properties, and the individual bees could no longer live on their own. A superorganism was born.

The population and density of bees had to reach a certain point for this to happen. Fifty thousand bees in a box the size of a mini-fridge allows for the rapid communication that the emergence of a superorganism requires. If those same bees were spread around the country and communicated by mailing each other postcards, well, that's not really going to do the trick, is it?

The situation with Agora is exactly the same. There has to be a certain number and concentration of humans for a superorganism to emerge, and we call that group of people a city. Cities are Agora's functional units.

But hold on a minute. Cities? They've come up several times already, but why? What does it mean that they are the functional units? If Agora is a planetary superorganism, what are cities?

There is a fascinating behavior of some ants that weighs directly on this question. Some species, when introduced into new areas that lack predators, form what are called supercolonies. They are gigantic, measuring thousands of square miles, and they relentlessly expand. One day you might see a mound, then return a month later and see more mounds have emerged nearby, perhaps hundreds of yards away. These are the same family of ants, which are increasing their territory, driven by a sort of—pardon the pun—Mantifest Destiny. Mile by mile they spread, crossing deserts, swamps, forests, and rivers. There is really no stopping them.

Now, say you took an ant from that first mound you saw and dropped her on one of the nearby mounds. How do you think those ants would react? Normally, she would be as dead as fried chicken in a matter of seconds, but these ants recognize her as kin and she starts getting to work in their mound. This peaceful mixing of ants from separated nests is called unicoloniality.

While it is somewhat surprising that the next-door neighbor ants get along with each other, it isn't astonishing. But this is: Say you took an ant from that first mound and carried her ten miles, or even a hundred miles, to another mound of the same supercolony, not the one next door. What do you suppose happens? Nothing. She's accepted as one of their own. This is unexpected because while these ants are technically related, they are something like fifty-third cousins forty-seven times removed.

Each individual mound is a superorganism. That's really important, so hold on to that thought for a minute. How do we know they are each a superorganism? Because it is at that level where we see the emergent capabilities, and it is at that level—the mound—from which no ant can live apart. While they can change mounds, they cannot live moundless. Think of this like a blood transfusion. You can move blood between people, but blood cannot live very long outside a person.

So, is the unicolony—the aggregate of all of the mounds that the ants can move among—an even higher-level superorganism? No. Maybe someday it will be, but it isn't yet. We don't see any new emergent capabilities here, nor any specialization, nor communication, nor any information sharing. It's just a bunch of separate mounds that don't have anything to do with each other, not a new emergent creature. Additionally, a "part" of the unicolony—a single mound—could easily exist on its own, since it doesn't even interact with other mounds.

But you can see how this situation might change over the next million or so years. One cluster of nearby mounds might differentiate from each other just a bit. Because of their structure or location, one mound might be a bit better at raising broods, while another is near a more reliable food source. A flow of ants between the mounds could share information. Over evolutionary time, you might get emergent properties, and a new, higher-order superorganism might form.

Mapping all this to humans is straightforward. The earth is one big super-colony of humans concentrated in mounds called cities. Just as those ants could go work in a different mound, humans can move between cities reasonably easily. Of course, there are practical limitations here, of language, politics, nationality, and economics, but in broad strokes people are largely able to migrate between cities, and they do: roughly a billion people change cities in a given year.

Human unicoloniality is made possible by something called human universals. The idea originated with anthropologist Donald Brown, who wrote a book in 1991 called *Human Universals*, which listed hundreds of "features of culture, society, language, behavior, and psyche for which there are no known exception." A few examples: body adornment, dance, death rituals, dream interpretation, government, inheritance rules, jokes, metaphors, and units of time.

Think about that. There are hundreds of specific aspects about culture with no known exceptions. This means that when different populations of humans stumble upon each other, after perhaps ten thousand years of separation, they will have hundreds of things in common. Big things. It is on the basis of these universals that we can relate to each other. That, along with our shared history and cultural touch points from Levi's to Legos, allows disparate groups of humans to mesh together easily.

So, if a city is a "mound" and the whole of humanity a unicolony we can move within, is the unicolony of humans a superorganism? With ants, it wasn't; but with humans it is. That unicolony is what I call Agora.

But this came about only recently. If you read the accounts of Marco Polo, for instance, you get a clear picture of what true silos of unconnected human civilization look like. Europe and China had almost no knowledge of each other, let alone did they function as an integrated whole. They shared no information, had no communication, and each was its own superorganism. But in our modern era, with our rapid communication, integrated supply chains, world financial systems, and high levels of mobility, the collective world of all humans shows astonishing new emergent capabilities. What are these emergent capabilities? Well, look around you. Everything in the world today that didn't exist a few centuries ago was made by Agora.

Even people in cities separated by both distance and culture have an astonishing amount of shared knowledge. An unfathomable amount of communication, one of the hallmarks of a superorganism, occurs all around the globe without any regard for borders. When a new idea or invention happens somewhere in the world, it spreads across the planet in an instant. Regions of the world have specialized, with some growing wheat and others manufacturing goods. Collaborations across cultures are a form of planetwide cognition, and information flows like water around the globe. Because of all this, the parts—the cities—can't really survive on their own anymore. If the world were destroyed except for Toledo or Torino or Toronto, those cities would not be able to survive because they have become specialized parts of Agora. They can't grow all their own food, make their own medicine, pump their own oil, and everything else they would need to do to be a self-contained superorganism. Most people in the city would likely die unless the city could evolve new capabilities in a matter of weeks.

Returning to our earlier question, "Why cities?," why are they the functional units and not states or countries or continents? Nations and provinces are just legal constructs. Your state's borders could change tomorrow, and you might have to read about it in the paper to even know it happened. Continents are merely geological facts. A city, on the other hand, is a dense collection of a large population of humans. It's a human thing, not a legal or geological thing. One may quibble as to where its border ends, but the city is the only one that comes about—definitionally—by its population of humans. A country ceases to exist whenever people agree that it does. It need not even have a population to be a country. However, if everyone moved out of New York City tomorrow, there would no longer be a city there. And even if everyone decides to stop calling New York a city, well, it still is one.

In the archeological record, you can actually see the city emerge as a superorganism. Our oldest cites were just groups of dwellings that looked like amorphous blobs. They were just heaps of humans. Only later—and we dive into this in the next chapter—do we see streets, town centers, marketplaces, and the rest. We went from blobs of dwellings to grid-like layouts pretty quickly, and I think Agora emerged when that differentiation happened. The evolution of the city as I have just described it is astonishingly like the evolution of multicellular animals.

Have you ever wondered why there are hardly any city-states anymore? Aside from Singapore and Monaco, they really are gone. And yet that used to be the highest form of human organization. In fact, Aristotle said the city-state was the highest form possible, writing in *Politics*, "And when many villages so entirely join themselves together as in every respect to form but one society, that society is a city, and contains in itself, if I may so speak, *the end and perfection of government*" (emphasis added). The Greeks had no interest in being a nation. Sparta, Athens, Corinth, and the rest of the gang were happy to remain city-states. They were self-sufficient, nonspecialized superorganisms. Cities can, and do, coalesce into nations for political reasons, but these are largely artificial constructs and, as we saw after the fall of Rome, when the nation dissolves for whatever reason, the city reasserts itself as the basic organizational unit of humans.

So Agora's functional units are cities. A few centuries ago, cities were stand-alone superorganisms, little mini-Agoras, but they aren't anymore because they are no longer self-sufficient and they are all interconnected—communicating, sharing information, and the rest. And through this, new emergent capabilities come about. However, cities can still be thought of and even understood as superorganisms because they retain so many of the characteristics of the superorganisms they once were. But we must always keep in the back of our heads that they aren't superorganisms anymore since they've specialized and have thrown in their lot with something even bigger. So, just like in section I where we saw life banding together to create ever-higher levels of complexity and emergence as it progressed from single cells to superorganisms, we see exactly the same thing happening in humans, from the mammoth hunters to cities to Agora.

But all of this is quite abstract. Let's bring it down to earth. How exactly does a city function like a superorganism? That's our next stop.

CITIES

HOW DID CITIES START? Was it the same set of forces that caused the bees to become social insects? Basically, yes. A few thousand years ago, humans stopped hunting and gathering in bands of a hundred or so and started forming settlements, which later turned into permanent cities. The catalyst for this dramatic shift in lifestyle was the invention of agriculture. Speculative illustrations of these earliest cities—the sort of thing you might have seen in your social studies textbook long ago—usually show them surrounded by wheat fields gently blowing, with residents using their harvest to make bread. They did raise wheat, but one credible theory posits that they grew it not primarily for baking bread but for brewing beer. Another suggests that we reluctantly became urbanites because it was a safer mode of living in a world beset by climate change, growing populations, and frequent human migrations.

Whatever the reason, it must have been really compelling because it was such a dramatic change in lifestyle. For fifty thousand years, behaviorally modern humans evidently had no desire whatsoever to make permanent cities, until suddenly, for some still-unknown reason, they went city crazy. Today's nine-to-five commuter with an "I'd rather be fishing" bumper sticker inching his way to work on a congested highway must sit there trying to figure out what our ancestors thought was so stinkin' wonderful about city life. And yet, in trading our nomadic lifestyle for a sedentary one, we have gained vastly more than we have lost by almost any measure. While it is easy to be nostalgic about a past that no one alive has ever experienced, I will point out that once the dentist starts drilling, no one sitting in that chair pines away for the good old days before the invention of anesthesia.

What is a city? While they can vary widely in size and population, we do see common elements. Broadly, they are permanent, densely populated human settlements with a substantial number of inhabitants. They have demarcated boundaries, something akin to a cell's wall, and are discrete political units as well, usually with a mayor and an administrative hierarchy, which provide services, collect taxes, and enact and enforce laws.

Disagreements about which was our first city boil down to debates not about years, which we are pretty good at figuring out, but over populations, which are more of a guess. What makes a "mere heap" of humans a city? Is it a population threshold? Or a population density? Or is it more about the social structure behind it, or is it simply its permanence? There isn't agreement on any of these questions, and dictionary definitions never list definitive thresholds for them. Çatalhöyük and Jericho are often cited as the oldest cities, but in his book *Metropolis*, Ben Wilson categorizes those as mere villages and gives pride of place to Uruk, a Mesopotamian city on the Euphrates in present-day Iraq that reached its zenith about five thousand years ago. Uruk actually means "city," so for a thousand years it was just "The City." As Wilson writes, "With a population of between 50,000 and 80,000 and occupying three square miles, Uruk was the most densely populated place on the planet. Like an anthill, the city sat atop a mound created by generations' worth of activity, layers of garbage and discarded building materials creating a man-made acropolis . . . visible for miles."

City living had many appeals, just like today, that touched all aspects of life, including the financial, cultural, and social. Anonymity was available as well. In the city, no one asked you why you didn't make a sacrifice to the gods last week, nor did they care, nor frankly even notice. The very idea of privacy must have been born in the city, and that, combined with anonymity, meant that peddlers of vice found they could do brisk trade on city streets. Thus cities gained early reputations as dens of debauchery, which could only have added to their allure.

Cities brought plenty of bad, too. High population density coupled with frequent visitors from distant lands meant that cities were petri dishes for a variety of virulent pathogens, which could—and often did—turn into devastating plagues.

Poverty looked different in the city than it did in the country. Rural poor were less likely to go hungry because they grew their own food and had a support network, that is, their community. Thus, out in the countryside you had poor people aplenty, but until the city you didn't really have what the inscription on the base of the Statue of Liberty termed "your tired, your poor, Your huddled masses yearning to breathe free, The wretched refuse of your teeming shore."

Despite all these downsides, the belief in the financial opportunities of the city was an irresistible lure that attracted people in droves, well beyond the carrying capacity of the city itself, often resulting in what today we call urban slums. Today, one in eight people—a billion individuals in total—live in such informal settlements, hoping to get just a tiny bit of the prosperity the city exudes.

Over time, wealth accumulated in cities, meaning they needed walls. Walls needed to be defended, so we invented the idea of the standing military and, with it, organized warfare. Cities needed laborers to build the walls, along with temples and palaces to appease both gods and kings. To accomplish all of this, institutional slavery was invented. Back in hunter-gatherer days, human enslavement probably didn't even occur to anyone. What would be the point? There was no wealth to accumulate beyond that day's haul of berries and grubworms.

As society urbanized, variations in wealth began to appear, brought about by variations in, well, luck. Biology capriciously bestowed on some people advantages in ability, health, appearance, and intelligence, while denying them to others. One could get a boost by being born first, or being born male, or even being born under propitious stars. The vicissitudes of daily life further separated people into proverbial wheat and chaff, with some falling ill and others enjoying good health. Since wealth could be accumulated and passed down to one's progeny, over time those variations compounded upon each other, resulting in a stratified society. The wealthy and powerful were motivated to protect their wealth and power, and—as a reward to themselves for all their hard work and foresight—reserved for themselves privileges and prerogatives forbidden to others. The Aztecs, for example, would restrict the smelling of certain flowers to the upper class.

This dense concentration of people from all backgrounds, interacting and planning together, took Agora to a whole new level. This wasn't just a band of hunters taking down a mammoth. The density of people combined with the rapidity of their communications would make the city the birthplace of most new ideas. It has been suggested that the stimulating effects of coffee helped usher in the Enlightenment and the Scientific Revolution, but I would be remiss if I didn't point out that during Prohibition in the United States, patent applications in formerly wet counties fell by one-seventh. So beer—which is so old that it is literally prehistoric—might be the unsung hero of the history of ideas, both good and bad.

Later, cities would have infrastructure designed entirely to foster and promote such communication. In the Victorian era, for instance, residents of London received mail twelve times a day, and etiquette demanded that you answer inquiries at the next delivery, an hour later. It was like email.

Bob Metcalfe, the inventor of Ethernet, coined something we call Metcalfe's Law, which states that the value of a network goes up exponentially as the number of users goes up linearly. Imagine the early days of the telephone. If there were just two people with phones in a town, then the number of possible conversations is just one. But if you simply double the number of people with phones to four, that is, A, B, C, and D, then the number of possible conversations doesn't just double, but goes up sixfold (AB, AC, AD, BC, BD, CD). Scale that up to a town of ten thousand people, and you will find fifty million different conversations that can happen. That's starting to look like neurons and synapses, and therein is the power of the city to create new ideas. That is Agora thinking.

In *Metropolis*, Ben Wilson sums it up well: "This rapid series of inventions and refinements was possible when humans clustered in a dense, competitive environment. Innovation begat innovation. The high temperature of the potters' kiln was used for experiments in metallurgy and chemical processes."

But the real power of the city doesn't come from communication alone but in combination with specialization, that is, the division of labor. Estimates of how much that increases human productivity are all over the map, but it is probably on the order of a thousandfold. The pin factory story referenced earlier from Adam Smith's *The Wealth of Nations*—allegedly an actual empirical example—demonstrated a rise in productivity of roughly that magnitude after specialization was introduced.

While social insects do amazing things with just a few categories of occupations, our level of specialization defies the imagination. There are literally hundreds of thousands of jobs. The US Bureau of Labor Statistics tracks the financial outlook for about ten thousand different ones, which include railroad detective, dog bather, boiler coverer, spinning doffer, sedimentationist, spike machine feeder, ringmaster, stained glass installer, and the oddly specific relish maker. Interestingly—and I am not joking here—it doesn't list beekeeper but does list bookkeeper.

We've already explored the thousandfold advantage of specialization, but it's worth reflecting on it in the context of a city. A group of people is as smart as the aggregate knowledge of them all. So a band of hunter-gatherers who have largely the same knowledge isn't all that much more impressive than just one person. But a hundred people who all know different things is another entity altogether. The power of this is not just that we collectively know more, but that it frees us all up to know considerably less. There are a thousand skills I never have to acquire because someone else in town knows how to do them. We'll expand this cognitive division of labor shortly.

But even communication and division of labor aren't the whole story. Cities encode, and over time accumulate, information. I mean this in a literal sense. My family lives in an 1890s Victorian home with three fireplaces. Shortly after we moved in, I hired a locally renowned chimney guy to look at each of them to make sure that they were safe to use. As he chain-smoked Lucky Strikes in my new home, he took great care to show me all the things that went into the design of my fireplaces, especially the angle of their back walls, which he said had obviously been rebuilt according to the specifications of some such-and-such 1920s British coal report that had measured blah blah blah—this is where

I stopped listening. I mean, this is *the guy*, right? Either he is the best chimney man on the planet or a crazed pathological liar posing as the best chimney man on the planet. Either way, I didn't want to get on his bad side.

Think about it for a minute. The angle of the back walls of my fireplaces is encoding information. A whole lot of it, in fact. Much research and toil must have gone into that report my chimney guy was referencing, along with an equal amount of experimentation, trial and error, fieldwork, and all the rest. All of it just to determine the best angle to deflect heat into the home while sending all the smoke up the chimney.

Of course, bad information can be encoded as well, to deleterious effect. I am six foot four, and every single time I walk downstairs, I have to tilt my head to the side to not bonk it on the overhang between the floors. Our staircase is encoding information about the height of the humans it expects to traverse it and, at least as far as I am concerned, they undershot it by a couple of inches.

In 1943, Winston Churchill addressed the House of Lords on the reconstruction of the House of Commons building, which had been bombed two years earlier. He was advocating for rebuilding it exactly as it was before, saying, "We shape our buildings, and afterward our buildings shape us." I think Churchill was referring to the degree to which entire buildings encode information, just like my fireplace. Vast amounts of information go into the layout, the signage, the routes between related areas, the size of the rooms matched to their purposes, and even the thickness of the walls and the privacy they provide their occupants—or fail to. There are the locations of the doors, the directions in which they open, whether they lock, and the height of their knobs, along with the windows and their various sizes, locations, and purposes. The placement of the emergency exits, the bathrooms, the water fountains, and all the rest should ideally fit their occupants, exactly the same way that flowers and bees coevolved to perfectly fit together.

Expanding on that, think about how much more information a city encodes in its streets, awnings, drainage, parks, sidewalks, roads, monuments, and the locations of trash cans, fire hydrants, streetlights, parking spaces, hospitals, and everything else. Try to imagine if every nook and cranny of a big city encoded as much information as that back wall of my fireplace. That's pretty close to the reality of the situation.

With all that background, let's consider how a city behaves like a superorganism. Let's look at one example, New York, and just one part of it, the twenty-three-square-mile island of Manhattan. I choose it specifically because it is an island, linked to the rest of the world by a few dozen bridges, tunnels, and ports. This makes it a much more discrete unit than the five thousand square miles of Tokyo. Within the dense hustle and bustle of Manhattan with four hundred years of information encoded in it, we can clearly see a how a superorganism emerged.

Let's start with food. Every day, the people of Manhattan eat ten thousand tons of food, most of which is brought onto the island on some of the 125,000 trucks that make the daily journey. That food is delivered to roughly forty thousand outlets, including ten thousand restaurants.

Who exactly is in charge of coordinating all of that? How do they make sure that there is the right amount of flour for the bagels and pizzas? How do they guarantee that each of the million items available across the menus of those ten thousand restaurants can be made every day? You know the answer: No one is in charge of it. It's like a beehive—all the workers just do their jobs, and in so doing the hive thrives. The ten thousand restaurant managers in Manhattan each place orders with their suppliers, who in turn aggregate those orders, up the line. All of that is then balanced with the variable supply, which hinges on factors like how many cod were caught in the Chesapeake Bay the day before or whether there had recently been a cold snap in the Midwest that blighted some crop.

Each bee has a few simple algorithms that guide its behavior, and the combined actions of each of those bees doing its own thing makes the superorganism. We are exactly like that. But because we specialize so much more—recall those ten thousand job titles—the algorithms that we use are highly specialized as well. I personally have no idea how much food a restaurant should order because those aren't the algorithms I have, but it's second nature to a restaurateur.

And imagine the scale of it all: *ten thousand tons* of food. Every single day. It all has to be made, inspected, boxed, labeled, loaded, trucked, delivered, and unloaded. Every. Single. Day. And that's just one thing, food. Who is in charge of telling taxis and Ubers where they should hang out for rides? Who makes certain that all the buildings get cleaned? Who is responsible for making sure there are enough dog washers, boiler coverers, and relish makers? No one. It is millions of people in ten thousand specialized jobs, each operating on a few specialized algorithms. All of this comes together to make Agora, the emergent whole, and this is the way by which all of us are able to live in a world with a plethora of clean dogs, well-covered boilers, and infrequent relish shortages.

The humbling thing is that humans could neither design Manhattan nor manage it. The reason we can't deliberately encode all that information is not only that there is no one with all the information, but that the information that would be needed cannot be anticipated. It can only be encoded over time by the interactions of the parts of the superorganism, that is, through experience—through learning.

Let's look at a specific example. Not some big, dramatic urban project, but one of the mundane things that happens thousands of times a day in a big city. Imagine a small Midtown restaurant that has a loading dock in the back. The restaurant manager discovers that the dock is one or two inches too high relative to the lift gate of some of the delivery trucks, such that the back of the trucks can't fully open when they are docked there. The manager consults with the trucks' drivers, and they determine that a few two-by-fours laid side by side across the driveway for the back wheels of the truck to rest on will raise it an adequate amount when it is docked. So they do that. There is now one more piece of data encoded in that city, adding to the infinitude that have accumulated over the centuries. This is the coevolution of the superorganism with its parts: the city and its people.

So cities are hives, the physical manifestations of Agora. We, the parts, act autonomously, running our algorithms, and try to maintain homeostasis in that complex system, and that work never stops. There's a reason they call New York "the city that never sleeps."

However, while it may not sleep, if it really is a superorganism, then it must be alive. If it is alive, then it must exhibit the characteristics of all living things. That means that it must have a beginning and that it must be able to die. It may be like a lobster and not have a fixed life span, but it must at least be *able* to die.

Recall how carefully the swarming bees choose the site of their new home. When humans form cities, their sites are (usually) picked with similar care. Even the earliest cities had a laundry list of requirements: The site needed to be defensible, have building materials nearby, abut land fertile for agriculture, and, most importantly, have access to water. Water was, is, and always will be the most essential requirement for life. It was needed to drink, obviously, but also for livestock, irrigation, communal baths, and washing away sewage. Thus, early cities were built next to rivers. Las Vegas would never have made the cut.

Cities are highly resilient, but they can and do die. The ruins of ancient cities that litter the earth bear witness to this. How do cities die? One way is climate change of one kind or another that makes the location less ideal, by, perhaps, making it so that a key staple crop can no longer be grown. Cities have died because rivers have altered course or wells have dried up. But "death" may not be the right word. Perhaps "hibernation" is more apt, for when conditions change again a few centuries or millennia later, people instantly flock to the old location and start building again. Social insects do likewise, since populating an abandoned nest or hive is easier than making one from scratch.

Sometimes in war, cities are destroyed as collateral damage or are even deliberately obliterated. History has more than one example of a city that was leveled, then a river redirected over the location, and the surrounding fields blighted by being sown with salt. However, more often than not, we don't know why any particular ancient city was abandoned.

Cities can also expand and contract in dramatic ways. Rome hit the amazing population of a million people around the time of Christ, but just a few centuries later dipped to just 3 percent of that. And although it is unusual, instead of abandoning cities when adverse conditions occur, cities have occasionally been literally moved, such as in 1919 when the two hundred buildings in Hibbing, Minnesota, were painstakingly hauled west two solid miles.

Not long ago, in the early days of the internet, it was trendy to predict the end of cities. The thinking was, "With technology, people can live and work anywhere, not just in a city. And who wants to deal with the city's crowds, traffic, and high cost of living? Instead, people can live in quaint coastal towns or, if they prefer, in the middle of nowhere. After all, who in their right mind would choose Chicago over Carmel?" This was a perfectly reasonable prediction, but as you know, it didn't shake out that way. For many, or even most, the appeal of cities turned out to more than offset their higher cost and inconvenience, and thus the urbanization of humanity has continued. As Ben Wilson wrote in *Metropolis*, "Today the world's urban population grew by close to 200,000 people. It will do the same tomorrow, and the next day, and so on into the future. Two-thirds of humanity will live in cities by 2050. We are witnessing the greatest migration of history, the culmination of a 6,000-year process . . ."

This is for the best, I am sure. Cities are the greenest way we know how to live. It isn't Times Square that sucks up the resources, but suburbia with its endless sprawl of manicured lawns with unquenchable thirst. The density of cities is what makes them efficient. The ecological footprint of the city is vastly larger than the city itself, of course, because it includes the land needed to grow the food, raise the animals, generate the power, and so forth. But still, it's the best we can do with the tools we have.

What about people who don't live in a city? If we are talking about Grizzly Adams or someone, well, that's the next chapter. But if we are just talking about people in the suburbs, or on a houseboat or who live in Crabapple Cove, Maine, or someplace similar, well, they all are part of the periphery of the city. They still are all connected to the information sources, still benefit from the innovations, still communicate with the city, and all the rest. Beehives have bees whose only job is to hang out at the entrance to the hive and keep out the riffraff. They never see the queen or do the dance or tend the brood, but they are still part of the hive, just not fully immersed in it.

What is the net of all this? Yes, cities are born, and they do self-create. They do not have a preset life span, but they will eventually die. All of this is consistent with their being part of a superorganism. There is another interesting proof point. The fact that we do a poor job planning cities from the top down is a strong argument for Agora as a superorganism. Bees would probably design a terrible hive; instead, they build one from scratch, organically, with its design influenced by its environment and modes of use, which is the same way you make a city that can coevolve with its inhabitants. It really is all the same. This is also why no two cities are remotely alike, because they are both part of the growing, changing, learning superorganism. Ant mounds behave in a similar fashion: Each one is quite different from the others, in both form and temperament, even though they are all created using the same simple algorithms.

Another characteristic of superorganisms is that their parts can't survive apart from them. Does that apply to humans?

CAN WE SURVIVE
ON OUR OWN?

HUNTER-GATHERERS WERE JACKS AND Jills of all trades, the original Renaissance men and women. However, their knowledge must have been a mile wide and an inch deep. They really were jacks of all trades but masters of none. I suspect there was no one alive in the Paleolithic era who could make relish anywhere near as well as one of today's relish makers. But if someone from the last ice age along with a modern relish maker were dropped on a desert island for some reality TV show, I know who I would put my five bucks on.

Our specialization has served us well. Before cities, the human population was thought to be well under a million. Even when Uruk was at its height, world population was probably only around eight million. The story of how we got from eight million to eight billion is the story of how specialization empowered us to become a thousand times more than we formerly were.

To achieve the incredible increases in productivity that specialization makes possible, each part of the superorganism has to focus on a narrow range of things that it does well, to the neglect of much else. As society grows more complex, each of us knows how to do a smaller and smaller percentage of everything needed to maintain it. Maybe in a band of hunter-gatherers, any individual could do 80 percent of everything needed to survive, but each of us, as members of a complex society, may only be able to do 1 percent.

So, have humans done what all other creatures who form superorganisms do? That is, become so specialized that we can't survive alone anymore? By "alone" I mean the proverbial "dropped on a desert island" scenario where we have nothing but the clothes on our backs and the knowledge in our heads.

Given this setup—modeled after a single bee being separated from the hive—I think we would all fare poorly. This is for two reasons, both relating to specialization. The first is what you normally think of with the division of labor: We lack the broad range of skills that we would need to survive. The second reason is that in addition to the "regular" division of labor, there is a cognitive version of it as well. It turns out for us, thinking is now a communal activity, and so we can no longer do that on our own either.

We are going to look at these one at a time. But before we dive in, I want to point out that while it would be easy to find this chapter discouraging, I offer a different interpretation. The overarching idea to keep in mind is not how incompetent we are. We obviously aren't. The big idea here is coevolution, which we covered earlier. That's where the bee and the flower evolve with each other and form a symbiosis of sorts. Individually, we haven't regressed; we've simply coevolved to live in community together. If we are removed from that community, there is no shame in the fact most of us couldn't survive, any more than a bee should be ashamed it cannot survive without flowers.

So how many of us could muddle through by ourselves? Obviously, we aren't trafficking in absolutes here. The portion of people who could survive on their own is neither zero percent nor a hundred. Humans are so amazing, so resourceful, and so intelligent that a few are bound to make it through. This wouldn't disprove Agora as a superorganism because as we will see, most of us wouldn't make it, and the few that would wouldn't have an easy time of it.

Let's start with a hypothetical one hundred random Americans, a group I can find good data on. Let's look at some numbers that might bear on their ability to survive the physical rigors of living alone. We aren't even talking about lack of knowledge yet, just whether people could physically survive separated from civilization. According to the CDC's numbers, eleven of our one hundred have severe mobility issues, five are blind or nearly so, and three have difficulty with self-care. All of those folks would have added challenges right out of the gate, but they still might make it. There are people who need life-sustaining medicines: the three that need insulin, the two on post–heart-attack medication, the one on post-stroke blood thinners; and none of those would be available any more. In addition, about a quarter of Americans take three or more prescription drugs a month, but we don't know how many of those they couldn't live without. Finally, some portion of our hundred are at extreme ends of the age spectrum. Most centenarians and toddlers would be hard-pressed to survive alone.

What is the net of all of this? It's hard to say exactly, but just at a gut level, let's say that half of Americans wouldn't make it very long based on their age and health alone. That actually speaks incredibly well of us as a society. One of the best things about modernity is how we are able to help people at the margins of health and wellness not just live but also participate in the daily life of our culture.

How would the other fifty fare? Let's assume they are scattered randomly around the world, and where they landed would have a great deal of influence on whether they survived. If in a desert, well, game over before it starts. Same with the far north and south. While tropical islands have their romantic appeal, life there wouldn't consist of long walks on the beach at sunset sipping mai tais. So let's remove half of our fifty for ending up in less-than-hospitable locales.

The twenty-five still left will not have it easy. After all, they don't get to take a pickup truck of equipment with them. That's a piece of Agora. As we've discussed, that's billions of person-years of innovation and the combined knowledge of millions of people. That's not living apart from Agora; that's not surviving on your own. We might as well let them take a camper and have groceries delivered twice a week. We're talking about a scenario where people are separated from all of that. This is no *Gilligan's Island*, where they each happen to have brought a wide variety of handy items, which is odd for a three-hour tour. A three-hour tour.

So, of the twenty-five who have made it over our first two hurdles, how many would know the following things that would be necessary for survival? Let's use you, the reader, as a proxy. Do you know all of this:

- The berry rule: 90 percent of yellow and white berries are poisonous, 50 percent of red ones are, and only 10 percent of black and blue ones are. And are you familiar with the technique to determine whether a particular berry is poisonous? Rub some on your arm first, wait a few hours, then a tiny piece on the inside of your cheek, and . . . well, it goes on, but even if you know it, it isn't foolproof.
- Don't eat if you don't have water. Especially don't eat protein without water, or you'll succumb to protein poisoning.
- Ahh, water. You need a gallon a day. If you found yourself in a tropical paradise, would you drink coconut milk? Well, you may have just signed your death warrant. It's a laxative, and if that's all you drink, you will quickly die of dehydration. Then there is water with parasites in it. Could you tell likely safe water from unsafe, assuming you had access to any at all? And don't forget to boil your water. Oh, wait . . . you don't have anything to boil it in.
- The correct way to prepare and eat cattails? Or whether you should eat plants with milky sap? Or waxy leaves? How about wild onions? In theory they are fine, but there are toxic plants almost identical to them. Can you tell them apart?
- Fishing probably isn't a good use of your time since you would probably spend more calories trying to catch a fish than you would get by eating it. And would you eat it raw? There are those parasites again. Even sushi must be frozen for fifteen hours at –31 degrees Fahrenheit to kill them. Serving raw fish that hasn't been frozen is illegal in the United States and elsewhere.
- Then realize this will all likely change with the seasons. How will you overwinter? Or endure a blistering summer?

Scurvy and malnutrition are just waiting for you to slip up, along with starvation. Maybe you've heard the story of Chris McCandless, a twenty-four-year-old who decided to be a nomad and live off the land in the Alaskan bush. He set out in April 1992 and happened upon an abandoned bus, which he decided to live in. He is suspected of raiding some empty cabins for supplies, but, in spite of this, he died of starvation by August, weighing just sixty-seven pounds. After just four months. Summer months at that.

Then, you better hope you don't cut yourself, because if the cut becomes infected, that's almost a death sentence without antibiotics. Don't get appendicitis, or gallstones, or kidney stones, or anything that would require surgery. Or a toothache that needs a root canal. Let's hope you aren't near predators or venomous animals. Cross your fingers that there's never bad weather.

We haven't even touched on building a shelter or maintaining your mental health. Maybe you would have a volleyball named Wilson to talk to, but that would be a best-case scenario. Suicide would always be lurking in your mind as an option, but there isn't even a quick and easy way to do that, let alone a painless one.

A 2021 survey of two thousand Americans revealed that only one in seven are very confident they could start a fire with flint and steel, which you probably won't have. Only one in seven are confident they can identify edible food. Only a third could spot poison ivy in a photo. Collectively, on average, they thought they could survive sixteen days alone in the wilderness. I admire their optimism.

My guess is that few of us could actually last in the wilderness. Even most survival experts would perish without appropriate gear. If we bent the rules and let everyone have a backpack full of Agora-created gadgets and books, I think they only postpone the inevitable.

Certainly, we hear amazing stories of survival against the odds. But that's because they are so rare. And the survivors always have some amount of equipment. Earlier, we looked at the Grafton castaways—you know, those guys who were stranded on an island and started brewing beer. But keep in mind that that was 1864, a time with less specialization, and they were all lifelong sailors who had lived in incredibly harsh conditions before. It wasn't even the first shipwreck for all of them. Finally, remember that we are talking about how many people can simply survive. We haven't even touched on thrive.

Sure, Bear Grylls might be able to eschew society and make it on his own in the Yukon with nothing but a paper clip, a slingshot, and a piece of beef jerky, but he's basically an honorary hunter-gatherer. The rest of us mere mortals have thrown in our lot with society and have specialized to such a degree that we can no longer function on our own.

But that's just the first of two reasons we can't survive on our own. Let's say a few people make it past all of these hurdles. What next? You might be thinking, "Humans are a pretty smart bunch of folks. Sure, the few weeks would be really rough, but the setup of the problem is that they get to keep everything they know, and humans know a great deal. They'd be brewing beer in no time."

I wish that were all true, but I don't think it is. There's the second reason, the cognitive division of labor, which would doom even the intrepid few who've made it this far. Let's look at that.

Our "thing" as a species is that we are the smart ones. Our brains are often hailed as the most exquisite, complex things in the known universe, but that's a bit self-serving since they are the ones doing the hailing. But they do have a point. Our position as the most powerful creature on the planet is based not on our bodies but on our minds.

Yes, we are smart, undoubtedly, but that doesn't mean we have "general purpose" intellects. Instead, we have highly specialized ones. It isn't that we're dim bulbs, rather that there is a cognitive version of the division of labor. Not only can we *do* just a narrow range of tasks, but we only *know* a narrow range of things as well. I know it doesn't feel like that, but each of us knows less than we think we do.

I recall an old *Night Court* episode where the bailiff is struck with appendicitis and is taken to the ER. The only doctor on hand is a podiatrist, who bristles at the suggestion that he is less qualified than a regular doctor. When the bailiff lets out another agonizing scream, the podiatrist suggests that "it might be his shoes." That's each of us.

Polls that show how ignorant we are make good headlines. You've probably seen them, too, how only half of Americans know the sun is a star, or how more people can name the Three Stooges than three justices on the Supreme Court. But that is all just clickbait, really, and it's irrelevant. What does matter are all the things we think we know that we actually don't.

In 2002, Leonid Rozenblit and Frank Keil, both of Yale, published a paper called "The Misunderstood Limits of Folk Science: An Illusion of Explanatory

Depth." In it, the authors' opening sentence sums up their findings well: "People feel they understand complex phenomena with far greater precision, coherence, and depth than they really do." It is a pretty humbling paper to read. Spread throughout its forty-two pages are the results of a range of experiments that show that while people think they understand a certain object, such as a speedometer, a zipper, a piano key, a flush toilet, a cylinder lock, a helicopter, a quartz watch, and a sewing machine, when asked to explain it in detail, the subjects faltered.

Four years later, Rebecca Lawson at the University of Liverpool added to these findings by showing that even people's understanding of objects they are familiar with is similarly impaired. She asked subjects to self-score their knowledge of bicycles, and then she gave them a bare-bones drawing of a bicycle frame and asked them to correctly place the chain, the pedals, and any parts of the frame that were missing. Forty percent made errors, more or less, across all the levels of reported bicycle knowledge, even the self-described experts.

But it isn't just technology that we think we understand more than we do. We also overestimate our knowledge of natural phenomena such as rainbows and tides, as well as economics, politics, history, and, well, pretty much every other sphere of life.

What's going on here? Are we that deluded about what we know? Philip Fernbach, a professor at the Leeds School of Business at the University of Colorado and coauthor of the book *The Knowledge Illusion*, suggested in his TEDx talk, "As individuals, we do not know enough to justify almost anything we believe . . . [L]et's think about a couple of really obvious facts. We all believe that the earth revolves around the sun . . . But on what basis? Can you explain the astronomical observations that support that belief? I know I can't."

He uses the example of flat-earthers and compares them to the rest of us, saying that "they're not all that different than you and I." He suggests that when we meet someone who holds a view that we regard with contempt, we should realize, "They did not arrive at their position via a rational process of evidence evaluation and they don't understand the issue in depth. But neither do you!"

This is a biting truth, to be sure. But upon reflection, we shouldn't be embarrassed by it. It is just another form of the division of labor—a cognitive form. Just as, as mentioned earlier, it is a gift to be able to use technology without understanding it, it is also a gift that we can benefit from the knowledge of others as if it were our own.

Maybe we have all imagined what it would be like to travel back in time and amaze the locals with the astonishing things we knew how to do. Until, of course, we realized we don't know how to do anything. If I could somehow craft a laser pointer, I could start a pretty compelling religion. But obviously I couldn't. The best I could probably muster would be a mediocre flyswatter. There are a few things I would know that would be useful, like how you should boil water to kill the germs in it, but I sure wouldn't be making a light bulb or even a pencil. In fact, depending on what time period I went back to, there is a good chance that I would be more of a liability than an asset because I wouldn't know how to navigate that world.

But if we know so little, how do we survive at all, let alone thrive? The superorganism of us is able to "think" by using all of our collected knowledge. As Fernbach said, "Thinking is a social process. Rather than happening inside your head, it emerges from your interactions with those around you. People are a little more like bees than we often realize."

The challenge is that we don't differentiate what we know firsthand, which is precious little, from what we know second- or thirdhand. I've been to Hawaii so I know it exists, but I haven't been to Alaska, yet I am equally confident that it, too, exists. We all believe, at some level, in the transitive power of knowledge. One hypothetical example: Jill, who is an intern at NASA, tells you something about the moon you didn't know, and so you are inclined to regard it as a fact and confidently pass it along as such. But of course, Jill hasn't been to the moon, so she only got that knowledge from someone else that she trusts. Perhaps it eventually tracks back to something Buzz Aldrin was once overheard saying at a Christmas party, but maybe not. It turns out that for everyone who truly knows something, there are others, perhaps thousands, whose only basis for belief in it is that someone they trusted said it, which they in turn heard from someone they trusted, who got it on good authority from someone they trusted.

We shouldn't feel bad about it, for imagine if it were otherwise, that you refused to learn from anyone else and had no access to the knowledge of others. Our communal thinking is a feature, not a bug. Turning to Fernbach one last time, he says, "On our own, none of us knows all that much, we don't have to. We each have our own little slice of expertise, and our minds are built to collaborate and to share knowledge, which allows us to pursue incredibly complex goals when none of us has anything approaching the knowledge to understand it all."

Are there other species that are social thinkers? Steven Sloman, a professor at Brown University and Fernbach's coauthor of *The Knowledge Illusion*, weighs in on this, stating in an interview that "some cognitive anthropologists have made a strong argument that human beings are the only animals—indeed, the only cognitive systems—capable of this kind of collaboration." He adds that when a parent and child are doing something together, like building a sand castle, "They are literally sharing thoughts: they are pursuing a common end result and doing so with knowledge that they hold in common . . . If one runs into a problem, the other might help. This requires that they understand that they share a goal."

As social thinkers, we are all part of the cognitive division of labor. Society can't work any other way. This has proven enormously beneficial to society—certainly as much as the occupational division of labor—but it has a few downsides, mainly around how hard it is to dislodge erroneous beliefs. No one wants to find out that the facts they have built their life around are false. Cue the flat-earthers. Even if you don't know a flat-earther, you probably have had the experience of discovering—perhaps during a presidential election year—that many of your friends and relatives are, in fact, idiots, and you may even struggle to understand why otherwise good and intelligent people could bring themselves to support *that* candidate.

We don't quite do all cognitive work communally. John Steinbeck wrote, "Our species is the only creative species, and it has only one creative instrument, the individual mind and spirit of a man. Nothing was ever created by two men. There are no good collaborations, whether in music, in art, in poetry, in mathematics, in philosophy." Steinbeck overstates this assertion to the point of absurdity. Not only are complex creative efforts, such as movies, true ensemble works, but I can think of dozens of creatives that are only known by their associations with others: Rodgers & Hammerstein, Ethan & Joel Coen, Gilbert & Sullivan, Laurel & Hardy, Penn & Teller, to name just a few. (A piece of trivia: In movie credits and elsewhere, the use of an ampersand to connect two names denotes a collaboration, while an "and" means they didn't work together, such as when one scriptwriter is fired and replaced with another, but they both still had a hand in the final work.) But I leave this quote in because there is more than a germ of truth in what he is saying: Individuals are able to be creative entirely on their own, which I think is wonderful, because it suggests to me that we aren't merely a committee of a species, and that there is something almost magical within an individual human mind capable of true creation.

That aside, virtually everything in the real world is a collaboration and requires group thinking. You can't have modern society without this. Forget putting someone on the moon; no one can even make a pencil without a lot of help. We each have only a bit of the knowledge needed to build our world, and we think communally as well. That's all OK, and it has served us well, but it does seem to support the proposition that we cannot survive alone anymore, that we are part of an integrated whole.

Man, stacking all that stuff together like that sure makes us sound like a bunch of sad sacks. We can do very little, we know very little, and we cannot even think on our own anymore. But again, all of that just means that we have evolved to function as a part of a superorganism, of Agora, not that we are incompetent.

A final aspect of superorganisms is that they demand uniformity and conformity from their parts. And yet humans are famously varied. Or are they? Let's have a look.

CONFORMITY

ONE OF THE BREAKTHROUGHS in the Industrial Revolution was the idea of interchangeable identical parts. To us it is old hat, but it was a big deal at the time, and it helped make manufacturing at scale possible, which simultaneously increased the quality of goods while decreasing their price. But for it to work, the parts really had to be identical, and if a particular part was even a little different from the other ones—even if it was actually "better"—it was counted as a defect, because it couldn't interact with the other parts seamlessly. So, ironically, better was bad, and exact duplication, even down to the flaws, was perfection.

Superorganisms are pretty rigid places, and conformity is required. If an ant starts acting weird, the other ants will kill it. It is pretty easy to see why. The parts of a superorganism are supposed to be interchangeable as well, and the organism is only able to function if the parts interact seamlessly. Anything out of the ordinary puts the whole group at risk, so ants ruthlessly cull any deviants.

Human individuality may be the most compelling argument against Agora as a living superorganism. We may simply be too varied and individualistic to create the kind of predictable order needed to form a superorganism. Or maybe not. Perhaps our differences are only cosmetic and we are actually largely the same. Let's explore this.

All human cultures allow for some amount of individuality, and most even celebrate it. We choose our clothes, our hobbies, our food, our religion, our jobs, and where we live. We listen to the music we like, watch the TV we enjoy, and decorate our personal spaces in accordance with our whims and fancies. We pick our politics, our friends, our passion projects, and the words we speak.

Society's tolerance of individuality has varied considerably over the ages. You don't have to wind the clock back too far to revisit a time when we had a narrower range of ways to express our unique personality. But today's modern world more and more at least claims to agree with author David Grayson, who wrote, "Commandment number one of any truly civilized society is this: Let people be different." Even in the most oppressive times in history, under the most restrictive autocrats we've ever had, there have always been ways, sometimes covertly, to express one's own distinctiveness.

Doesn't that mean we aren't the cogs in a machine after all? Regrettably, I don't think so. We encourage individuality, but only in limited, narrow ways. IfIdecidedtoforgotheuseofspacesinmywriting, my book sales would rightfully plummet. Poet e. e. cummings is popularly remembered for his rejection of capital letters, which is about as far as you are allowed to flout literary convention and still have a readership. Beyond that, there is little tolerance. Start concluding all your emails with ". . . in accordance with the prophecy" and see how that works out for you. But it isn't just what we say and write. Try walking down the street naked and see how far you get. Or if you're too bashful for that, just try skipping everywhere you go; the result will ultimately be the same. I've never seen a man with half a mustache—either the left or the right—which is quite odd in a world with hundreds of millions of mustaches. If you drive down the street in reverse, regardless of how well you do it, you will be stopped and given a field sobriety test.

But all of these are pretty egregious departures from social norms and are written with my tongue firmly in my cheek. None of us are really going to do those things. What we are about to see is that while certain types of conformity are absolutely necessary, society goes way beyond those so that even minor, insignificant deviations are not tolerated either.

Society relies on a certain amount of conformity. As such, many of our bedrock institutions are designed to promote homogeneity. Our public school system is a good example. It's designed to teach a range of basic knowledge and life skills that help people function in our complex society. That makes perfect sense, of course, and it fills an important role.

What's interesting about it, though, is how it is structured. Our K through 12 system is a nineteenth-century Prussian invention designed to produce factory workers for an industrializing world; so in pursuit of this, it was deliberately modeled after a factory: You have a manager—your teacher—who gives you work. If you have a question, you are supposed to raise your hand and ask your manager what to do. When you finish an assignment, your manager assesses whether or not you did a good job, and if you repeatedly do well, you will get an annual promotion. There is a bell that rings to tell you to eat lunch, to stop eating lunch, to change tasks, and finally, to go home. Thus, our schools themselves are factories, built to manufacture factory workers. There is nothing inherently wrong with this approach except that perhaps it is a bit outdated. The point is that a manufacturing mindset that views deviation as error is baked into the system, resulting in it producing largely homogeneous workers. Everyone is taught more or less the same things, and standardized tests are administered to measure how standardized the workers are.

Other institutions are designed to promote conformity as well. Militaries shave everyone's head—at least all men—mostly so they look alike. Soldiers march in lockstep to blend together as an undifferentiated whole, and while they break stride when crossing a bridge so as to not accidently hit its resonant frequency, it seems to be a practice only grudgingly tolerated. Armies don't generally have three or four different uniforms soldiers can choose between depending on their mood on a given day. Dress codes in general, be they for school or work, are designed to mute differences, and they seldom encourage attire that can be described as avant-garde or edgy. Prohibitions against unnatural hair colors, facial tattoos, and body modification are justified because those variations are deemed to be distractions.

In addition to requiring the parts to be alike, superorganisms also require that the algorithms that govern the actions of the parts must be alike as well.

Are ours? I think so. For instance, in most occupations, the goal is for all the workers to do their tasks the exact same way. Sure, there are exceptions. No one expects that two novelists separated from each other will write the same book. But in most situations, good job performance is defined as doing a job the way you were trained to, that is, just like everyone else. When you go into the bank to cash a check, it doesn't really matter which teller you get because they are all trained to do the job the same way. It's the same with dentists, waiters, lawyers, electricians, and, well, about everyone else.

Even our most complex jobs—take brain surgeon, for instance—are also performed in carefully prescribed ways. Variation from the accepted protocol can actually be a criminal act, and it is certainly a legal liability. Certifications in jobs are granted based on one's ability to perform actions like everyone else. And that's the way we all want it. Would you want a brain surgeon who said, "Board certified? Ha! Who needs that? I'm self-taught." There's a reason that doctors hang diplomas in their offices.

Of course, in many areas of life there is good reason to require conformity. We are able to build skyscrapers only because a wide variety of workers with diverse backgrounds are able to mesh together and work toward a common goal. Even if the algorithms of two workers are a bit different, the fact that they have spoken language as a "least common denominator" allows them to sync up their algorithms by just sharing a few sentences.

Many of society's rules are entirely arbitrary but exist for good reasons, namely so that we can all coexist. There's no particular reason we drive on one side of the road or the other, or why red means stop and green means go. The point is simply that we all need to do certain things the same way or there is chaos. That also makes perfect sense.

Sometimes, however, demonstrations of conformity are required as a sort of test of one's commitment to the collective. The pledge of allegiance in the US is one example. If a student closes it out with "That's all folks" in Porky Pig's voice, well, that's a week of detention right there. And that's just one of many examples. Military officers must be saluted, prelates' rings must be kissed, and all must rise for the judge or be held in contempt of court. So, even contempt is criminalized. Religions have creeds that all the faithful dutifully recite every week, often in unison in a monotone, and they don't allow you to make a few edits here and there for brevity or clarity. If you go to a baseball game and the person singing the national anthem makes what they regard as a few improvements to it, they will literally be booed, and this act of nonconformity will be covered in the media the next day.

Grammar is another example of required conformity that becomes arbitrary when it goes beyond any practical requirement for communication. No one argues that the sentence, "I ain't never gonna do that" isn't coherent, just that it contains multiple errors. As I typed it just now, even Microsoft Word passed judgment on me with its smug little red squiggles. And yet, there is no objective right and wrong, at least in English. While some grammatical rules do promote clarity of meaning, much of what we call correct grammar is merely arbitrary convention. We're told not to end a sentence with a preposition and to never split our infinitives. The only justification offered is that that's the way we do it. We don't tolerate variants in spelling either, even if they are improvements over poorly spelled words. Why does "knife" still start with a *k* and "psychology" with a *p*? OK, there is at least one word where we do tolerate spelling variations, and that's okay.

Even groups of nonconformists conform with each other in how they don't conform. Those who reject the dominant conventions of their society often become conformists in a subgroup, wearing similar clothes or performing identical activities, and they in turn are often intolerant of deviations from those conventions.

There are a thousand tiny forms of conformity that we all follow reflexively because everyone else does. These are incredibly small, subtle things. But if someone begins to flout them, they really are making a statement that they are outside of the society, not part of it. Even the rejection of the most insignificant of these norms can be seen as a rejection of the whole. In some groups, simply wearing a bow tie when everyone else is wearing a straight tie will be noticed and judged unfavorably. If you decide to try to bring back the monocle or wearing a formal cape to events, well, kiss your career goodbye. Who knows what other deviant behavior you are hiding? Even innocuous acts of individuality that affect no one else—say, uncommon dietary practices—marks a person as an unpredictable loose cannon.

And that's the key word: unpredictable. That's the cardinal sin in a superorganism. After all, an engine only hums when all the parts are doing exactly what they are supposed to. If they aren't performing like the other ones, then they need to be replaced, and that's exactly what we do as well. We fire or shun or exclude the people who don't conform, even in areas where their deviations don't matter. It's the ant world again: If an ant acts weird, it gets killed. We aren't that harsh. We allow for nonconformity, but only when we are all doing it together. Noam Chomsky expressed a sentiment along these lines when he wrote, "The smart way to keep people passive and obedient is to strictly limit the spectrum of acceptable opinion, but allow very lively debate within that spectrum."

We have evolved a distrust of nonconformist behavior in general, but we have not yet learned to distinguish between harmful and harmless versions of it. The result is that we live in a highly regimented society where the range of actions that are socially acceptable is tightly defined, because, well, that's what it takes to make the superorganism hum.

If we are in fact part of a superorganism, then the organism itself is subject to evolution and evolves to fill some niche. What would Agora's niche be? Or, put another way, "Why are we here?"

WHY ARE WE HERE?

WE EXPLORED EARLIER HOW superorganisms come into being. They begin as mutually beneficial partnerships and then coevolution takes it from there, leading to more specialization until one day a superorganism is formed. Thus, the initial purpose of the superorganism is solely to further the survival of the parts. Beehives, for instance, may well have formed to provide a safe haven for bees, and since the hive maintains a steady temperature, it provides an ideal home for its inhabitants.

Over time, the superorganism continues to evolve with its parts, becoming better suited to them, and they in turn evolve to help it. For example, beehives went on to evolve the ability to store large amounts of honey as a backup food source for the bees. In theory, the entire purpose of the superorganism can simply be to further the survival of its parts, but I don't know any case where this is true. What actually happens is that in addition to evolving with its parts, the superorganism also interacts with the larger world. It evolves in the context of its environment, and so as it evolves, it affects the world around it. It begins to fill its own ecological niche, influencing and being influenced by its surroundings. For example, the honey stored in beehives is a food source for other creatures besides the bees. This includes bears, honey badgers, bats, birds, and ants, to name but a few.

How do you spot the ecological niche that the superorganism fills? By studying its emergent properties. That's the wild card, the shiny new feature that affects its environment.

What then is the niche that Agora fills? What is its essential function? Well, what are its emergent properties? Answer: everything in the modern world. Everything we have today that we didn't have ten thousand years ago is due to Agora. Hold that thought for a minute.

We must seem like aliens to the other creatures on this planet. Our abilities, along with what we have built and accomplished, set us apart in a profound way from anything else alive. Have you ever wondered why it is just us? Shouldn't there be a whole host of creatures coming up behind us in terms of intelligence? A broad range of animals at every point on the developmental spectrum? Shouldn't there be some animals entering their Stone Age, and others their Bronze Age, and so forth? But nothing like that exists. Nothing even 1 percent of where humans are at. If we see a chimp using a long blade of grass to get termites to snack on, we are amazed that it can use a tool. But come on . . . really? A blade of grass? That's supposed to be impressive?

What's holding the other creatures back? I wrote a book about that in which I argued it was that those animals lack a knowledge of the future and the past, so they couldn't plan. This was because they lacked episodic memory, that is, specific memories of past events. Humans gained the ability to do those things by virtue of a fortuitous genetic mutation that gave us not just those capabilities but all of the other human distinctives, including language and creativity.

Others have suggested different reasons that we're alone in our capabilities. Perhaps it turns out that you need four things to be a smart, successful, technology-accumulating, culture-building species: a big brain, fine motor control to manipulate a wide variety of objects, a long life in which to accumulate knowledge, and, finally, societal living so that ideas and information can be shared and passed down. If you miss just one of those, it just doesn't jell. Octopi have the first two—big brains and motor control—but they only live a couple of years and are solitary. Dolphins lack digits to manipulate their surroundings, so even if a dolphin could conceive of the telegraph, it could never build one. We just happened to be the creatures that drew four aces in a row and won the game.

But these are not explanations. They only kick the can down the street. Regardless of the feature that makes us unique, the question remains unanswered: Why just us? Why doesn't evolution make more things with episodic memory or opposable thumbs or the ability to plan or whatever it is that makes the magic happen? Why just one species?

To answer that, let's switch gears for a minute. What if there is a giant planetwide superorganism such as Gaia or something else like it? A living, breathing creature. Something that emerges from the totality of all life on Earth, unlike Agora, which is solely a superorganism of humans. What would that planetary superorganism want?

I happen to think that there is such a biological entity, but if you are not convinced, then let me reframe the same question: If the earth is simply an immense nonliving system of water cycles and carbon cycles and so forth, all combined into a single übersystem, what would it want? But wait . . . can a system "want" something? Yes, in the sense that it can have conditions under which it functions best. Colloquially, we speak this way, the way the owner of a sports car might say that the car "wants" to burn premium gasoline. But in more formal language we would think of it this way: Systems have feedback loops that they use to try to keep key parameters within ideal operating conditions. We saw this earlier in the ways that the earth's geological systems self-regulate, to try to keep certain parameters within a range of ideal values. In living systems, we call this homeostasis, and the bees who work to heat and cool the hive are trying to keep it in homeostasis. The equivalent term for nonliving systems is "steady state," which they "try" to maintain. They "want" to maintain that.

So, back to our question, what would such a planetary system "want," whether it is living or nonliving? There is a straightforward answer to this. If it is alive, or even behaves like a living system, then it probably wants what all other living things want: to thrive and to perpetuate, that is, to live and reproduce. It sure wouldn't want to die—that's way outside of its ideal parameters.

Should it be worried about dying? Absolutely. The planet upon which we all reside is little more than a speck of dust floating alone in the unending darkness of an infinite universe. A lot could kill it. It is a statistical certainty that at some point in the future, the earth is going to get walloped by some dev-astating celestial object. The asteroid that killed the dinosaurs, the Chicxulub impactor they call it, was about the size of Mount Everest, which doesn't sound all that big, but since it was traveling at twenty miles per second, it packed quite a punch. Those illustrations you've likely seen of two or three dinosaurs looking up at the big fireball in the sky as if to say, "This can't be good . . ." would have never happened, given its speed. With that in mind, consider that there are tens of thousands of near-Earth objects of various sizes that cross our orbit, so eventually we will hit every one of them as well.

We would probably survive a similar impact, as would other life-forms, but it wouldn't be pretty. Such an impact will happen again, but whether it will be in a hundred million years or a week from Tuesday is anyone's guess. NASA takes this threat seriously enough to embark on multiple projects designed to protect us from such a disaster by intercepting threatening objects and altering their course by a tiny fraction of a degree.

In a paper called "The Resilience of Life to Astrophysical Events," authors David Sloan, Rafael Alves Batista, and Abraham Loeb attempt to determine "what cataclysmic event could lead to the annihilation of not just human life but also extremophiles, through the boiling of all water in Earth's oceans." They find that "although human life is somewhat fragile to nearby events, the resilience of Ecdysozoa . . . renders global sterilisation an unlikely event." While this is undoubtedly good news if you happen to be an Ecdysozoan—whatever that is—as a human I find it hard to take much comfort in their findings.

The authors analyze four existential planetary threats—supernovae, gamma-ray bursts, large asteroid impacts, and passing-by stars—and deter-mine that while catastrophes are common, life so pervades our planet, deep into its crust in fact, that it would be hard to kill it all. But such a disaster could set life back so far that it might never recover.

So it is reasonable to assume that a planetary superorganism would have a strong incentive to build some kind of defense against these threats, and the best way to do that—even with all the trouble we cause—is to spin up an intelligent species, perhaps one with opposable thumbs, long lives, social living, episodic memory, language, creativity, and planning skills who could come together to tackle that problem head-on, and would be highly motivated to do so as well. In short, the planetary superorganism would need us. Well, actually, it would need Agora.

To be clear, the planetary superorganism—Gaia or whatever—didn't make us. It isn't as if it fashioned the first humans from clay and breathed life into them. Rather it is to say that planetary-scale superorganisms that fail to evolve something like Agora to protect themselves from existential threats will inevitably die out. It also suggests something else a little unsettling—that superorganisms on planets where many different intelligent species evolve, each perhaps eventually becoming their own superorganism, also inevitably die out, probably by blowing themselves up.

You can see why we are a dangerous bet. Any species powerful enough to protect a planet from an asteroid is likely powerful enough to destroy that planet as well. We have deemed it a wise choice to manufacture tens of thousands of nuclear warheads—not merely just the few hundred that would be needed solely for a deterrent—and if that weren't enough, we've figured out how to engineer life-forms with life-destroying properties. The answer to why we are the only smart ones on the planet could be that intelligence is volatile substance. It might be that life simply doesn't need our kind of intelligence at all, at least on a day-to-day basis. Sure, we might be needed to deflect the occasional asteroid, but otherwise intelligence might well be a hindrance to the perpetuation of life. Life seems to do better without it, and definitely no worse.

So whether we are talking about Gaia or a nonliving system is immaterial. The net in either case is that on any given planet, if no intelligent life ever happens to evolve, then that planet will get whacked by an asteroid. Likewise, if there is a planet where lots of intelligent life evolves, they might inevitably blow themselves up. The Goldilocks number of intelligent species might well be one. Just enough to keep asteroids at bay, and hopefully not enough to destroy itself.

However, the planet doesn't just need an intelligent species to build a rocket to deflect an asteroid. It only works at scale. It needs billions and billions of intelligent creatures of that species to accomplish its goal. Imagine if the population of intelligent humans never reached more than a thousand. You can't have the kind of innovation we have had—and certainly not as quickly as we have had it—without lots of people. The planet needs a bunch of us, because not only does it needs lots of hands to do that work, but it also needs the lightning in a bottle that is an Isaac Newton or an Albert Einstein to periodically come along and propel us forward. However, it can't have too many of us either, because we would quickly use up all our resources and burn ourselves out.

So, for the planet to survive, it needs exactly one intelligent species that is able to rapidly grow its population to some critical threshold without eating up all its resources. In other words, it has to grow fast and big, but not too fast nor too big.

So, maybe out of a billion planets, five just happen to evolve life. On one, it never becomes intelligent and that planet is destroyed by a cosmic event. On the second one, a plethora of intelligent species emerge and one of them inevitably blows the planet up. On the third, a single intelligent species evolves but never really achieves the critical mass of population needed to produce a superorganism as powerful as Agora, so it gets taken out by an asteroid as well. On the fourth, a single intelligent species evolves but grows so big so quickly it burns itself out and vanishes. And then there is the fifth one, and we call that one Earth.

So that covers the first of the two things a planetary superorganism might want: to survive. It needs Agora for that. What about reproduction? How could a planetary superorganism reproduce? You might recall from earlier, one of the arguments against the Gaia hypothesis was that there is no way for favorable mutations to persist because Gaia can't reproduce. How do we answer this?

Earlier, we explored panspermia, the idea that our DNA came here through space and, having landed on this planet, took root and transformed it. There are two versions of it, and the idea I am about to offer applies to both. Directed panspermia—the idea that this was a deliberate act—seems like something an intelligent species might possibly want to do. Today NASA carefully sterilizes our planetary probes lest they inadvertently contaminate life in an alien ecosystem. But it isn't hard to imagine the opposite: that we decide to seed the universe with our life. This is not narcissism, rather the ful-fillment of an obligation. If intelligent life is vanishingly rare, then those few planets endowed with it have a duty to spread it. Other planets need an Agora to protect them as well.

The second form of panspermia says it could just be accidental and actually might happen all the time. When a medium-sized asteroid hits the earth, it can throw matter into space, which can certainly contain biological matter. Is it unreasonable to believe that such matter could drift around until it happens to land on a new planet where it can take root? Many plants on Earth use this strategy to reproduce. They throw their pollen into the wind with confidence that a few specks of it will land on another member of their species. Wouldn't it be amazing if it turned out that plants and planets reproduced the same way? Life may be spreading through the universe this way. It would scale quite quickly, even with the vast distances involved. One planet seeds another, then the two become four, then the four become eight, and so on. The cosmos may be dripping with this sort of biological matter, which would explain why life started on Earth immediately after forming.

So that is the net of all of this. Superorganisms that produce technology-using intelligent creatures like us to defend them from cosmic threats live on, and those that don't become lifeless wastelands. Likewise, superorganisms that produce star-faring peoples reproduce by sending those people out into the cos-mos. Planets or planetary superorganisms that don't do this die childless. It is just plain ole natural selection, only at a different scale. We are therefore the cosmic child of some other Agora, and our Agora will soon be able to reproduce as well. If this is correct, we will eventually find evidence of alien life, but it will be based on the same exact DNA that we have, and it will speak the language of GTCA.

So that's my thought on why we are here: to protect life. That's our ecolog-ical niche. What a wonderful job.

There is one last question I would like to address. Maybe you've asked yourself if planetary superorganisms are themselves just cells in a larger cosmic creature. If the universe is populated with many Agoras, are they also special-ized, and in aggregate do they form an even higher superorganism?

On the one hand, it doesn't seem like there could be all that many Agoras quite yet. The universe is quite young—just fourteen billionish years old—and since it took billions of years for our Agora to emerge and send people to space, it doesn't seem like there could have been too many generations before us.

On the other hand, Agora can have a near limitless number of siblings. DNA is famously easy to copy. It wouldn't take all that many billions of years to populate the cosmos that way. Even with today's primitive spacefaring technology, we could build probes that could go to every nook and cranny of the Milky Way, with its eighty-six billion stars, in just a few million years, which is essentially the blink of an eye at the time scales we are dealing with. But if this is true, we are returned to the Fermi paradox. If this is all happening, then why aren't DNA-bearing probes constantly zipping by our planet? Maybe they are, and we just don't see them.

Is it possible that Agora itself is a part of an even larger superorganism? Possible? Sure. But is there any evidence for that? Arguments in favor would be the sheer size of the universe and the number of planets in it. We don't know how many planets are in the Milky Way—it's certainly more than the number of neurons in your brain—but does that matter? Further, galaxies themselves are in clusters that could in turn be thought of as organs, and the total number of planets in the universe is likely about the same as the total number of cells in all humans—all eight billion—combined. In support of this idea is that if life is nested levels of complexity, then there is no reason it can't be one more level higher.

All that said, regrettably, there can't be such a creature. Unless we are missing something big about the nature of reality, nothing suggests that these planets across the universe are somehow communicating or are even able to influence one another. Space just may be too big for the kind of intense communication, specialization, and collaboration that seem to be required to create an emergent universe-sized superorganism. Superorganisms emerge from the density of their parts coupled with the rapidity of their communication. The universe is the exact opposite of dense. It averages just one atom per every hundred gallons of empty space.

Plus, the time scales are all wrong. The billions of years it took to birth our Agora is already a big chunk of the universe's total age, and so the cosmic creature of which Agora is but a cell would have a life span of trillions of years, and we don't think the universe will last that long. Such a creature would be possible, but not in our reality.

THE VERDICT

WHEN I STARTED WRITING this book, I didn't know what I thought about Agora. I introduced the idea and the name in the book *Stories, Dice, and Rocks That Think: How Humans Learned to See the Future—and Shape It*. At first, I envisioned it as a metaphor for how groups of people can do things that no individual could. But the idea really grabbed hold of me and I couldn't stop thinking about it, so I decided to write an entire book about Agora in which I tried to figure out what it was.

As I worked on the book and studied systems theory, I began to think of it as a real thing, as a system. Systems are made up of parts that interact with one another in prescribed and predictable ways. In doing so, they become a unified whole. The characteristics of a system are that they accept input, modify it in some way, produce output, and have mechanisms for constraining or limiting all of that. They also have a feedback mechanism that influences their future operation. Systems have boundaries that demarcate the limits of their influence.

What I learned is that they are everywhere. You can read back through that description of systems and, instead of a car or a clock, imagine the water cycle—how water evaporates, forms clouds, rains down, and so forth. Computer programs are systems, as are corporations. It is a versatile concept, and the fact that it describes so many things doesn't dilute the idea, but rather illustrates how pervasive systems are in our world. Agora seemed to fit all the requirements.

So at that point, I thought of Agora as a machine of sorts. I found this really interesting because if it is a machine, then it can be optimized, and it could also break. As I would read the news, I'd try to understand world events as if the whole of modernity were a system that was broken in places. I would ask myself why it wasn't functioning correctly. That is, what had failed?

William of Ockham, who lived about seven hundred years ago, was an English friar and philosopher. He coined what today we call Occam's razor. It really is one of the better razors. The idea is that the simplest explanation is most likely the true one. Or, put another way, the explanation with the fewest assumptions is most likely to be true. As I continued to write, I kept asking myself whether Agora was a simplifying assumption or an unneeded one. I found it to be quite useful for understanding the events of the world and human history in general.

The more I wrote, the more Agora felt like it was alive, like it was organic. It seemed to me to have something like a heartbeat, and I could see the systems of the world mapping as its various organs. I knew it was possible that it could be a creature because it was made of living parts, that is, us. So that led me to try to understand what constitutes a life-form, and that led me to studying superorganisms. Although as a beekeeper I was familiar with them in that context, I wanted to understand them in a general way. There is actually very little literature that does this; instead, most books are about ants or bees or the human body or Gaia, but few of them advance a general understanding of what makes a superorganism tick.

While I was immersed in all of this, I would walk around my native Austin, Texas, and that's when it really popped for me. I could see information encoded in the physical structures around me. I could see the city's inhabitants—the Uber drivers and restaurateurs I used in my examples—acting independently, giving rise to a superorganism, and I could spot emergent properties. That's when I came to believe that Agora is a living creature, and since at that point the idea of planetary superorganism was no longer a stretch, I came to understand Gaia as a living creature, an emergent being made up of all of the life in the world. Again, these are not religious beliefs but simply ones about biology. If cells can make bees, and bees can make beehives, well, there's no reason that would be the highest level possible. I think the levels go all the way up to the planetary level. And just because something is a far more complex form of life, that doesn't make it a god.

In fact, Agora shouldn't even be that much of a leap. Imagine if you lived centuries ago and read the findings of Robert Hooke, which we discussed at the beginning of this book. He was one of the scientists who popularized the idea that we are made of cells. Imagine that. You lived your whole life thinking it was just you in your body, and then this guy comes along and says you are made of countless small creatures. While that may have blown people's minds, it wasn't a cause for an existential crisis. Neither is Agora, for it is the same sort of mental shift, a reframing of how we understand the world.

Is Agora conscious? That's the tricky one, and of course no one knows. No one knows if anything other than themselves is. If our consciousness is an emergent phenomenon related to our complexity, then Agora, being even more complex than us, would likely be conscious. But complexity may not be what gives rise to consciousness. I don't even know if I *hope* Agora is conscious or not. It's enough for me to understand it as a living creature, one that we should all work to nurture and cherish, because, well, it's us. It isn't that there is "us" and there is "Agora," rather that we are Agora—a realization that gave me the title for this book.

That's my conclusion. What was yours? If you are so inclined, please drop me a note. My email address is byronreese@gmail.com and I would love to hear from you.

———

We have one final topic to cover in this book: Does any of this actually matter in a practical day-to-day sense? I think it does. I think understanding Agora helps us understand human history and that in turn gives us a way to speculate on our collective future. I also think it has profound implications on how we should live and interact with one another. We cover all of this in our final three chapters: "The Past," "The Present," and "The Future."

SECTION IV
MEANING

MEANING

THE PAST

WE'VE HAD WRITING FOR about five thousand years, and so that is as far back as our records go. Anything before that is termed prehistoric, and what happened then must be inferred from archeological remains. To help us try to comprehend those five millennia of recorded history, we divide it into eras and areas, such as "home life in colonial America" or "Ming Dynasty agricultural methods," and even as specific as those topics are, one can spend a lifetime studying them and only scratch the surface of what could be known. The challenge isn't that recorded history is all that long—it's just a blink of an eye, really—but that a hundred billion people have lived and died in that short time.

There's an alternative approach to understanding the past known as Big History that attempts to take it all in at once, to view it from thirty-five thousand feet, with the hope that from such a height, we can spot the overall themes of history and understand the forces that move it along.

This book tries to understand our history that way, as the biography of Agora. Agora was born just before we developed writing, and so while our history doesn't document its birth, it does contain an account of its childhood and beyond. And the overall theme of that story, I think, is progress.

Progress. What does that even mean? In the broadest terms, it means going from something bad to something better. This means it's a value judgment. But since people have different values, is there any common ground to be had?

Is there anything that we can universally regard as progress? I think so. As President Kennedy so eloquently put it, "In the final analysis, our most basic common link is that we all inhabit this small planet. We all breathe the same air. We all cherish our children's future. And we are all mortal."

Kennedy only touched on a few of these common values, but there are many more. Life is better than death, freedom is better than tyranny, knowledge is better than ignorance, health is better than sickness, peace is better than war, and kindness is better than cruelty. There are many things in the world that are bad, such as hunger, pain, misery, suffering, desperation, greed, and hatred, and many that are wonderful, including happiness, joy, health, friendship, hope, harmony, and love.

Are there those who would disagree with these values? Sure. A few. Perhaps 6 percent. I know, that's an oddly specific number, but here's how I got it: A few years back, Clemson University student Nathan Weaver was studying how to help turtles get across roads. As part of this effort, he placed a rubber turtle on the side of a road as if it were trying to cross, and saw that 6 percent of drivers deliberately swerved to hit it. A turtle. A harmless turtle.

Some would argue that the real number is even higher than that, and there were probably lots of people who just didn't notice the turtle or didn't want to damage their car. Some believe that evil people abound in great numbers, and that the Big History is actually not a story of progress, but an unending struggle between two opposing value systems, the one I describe above and its antithesis, one that sees hate as better than love. But the idea of Agora doesn't support this view, for a superorganism can't be at war with itself. We don't see anything in nature resembling a beehive where half the bees are plotting to kill the other half. Rather, we see a story of coevolution and cooperation.

I agree with Marcus Aurelius, who must have been thinking of something like Agora when he wrote the following passage nearly two thousand years ago in his book *Meditations*. I have paraphrased the text from my copy of George Long's 1910 translation of it, and I have included the original in the back of this book for your reference:

When you wake up in the morning, say to yourself, "Today I shall meet ungrateful, arrogant, deceitful, envious, unkind people. They are this way because they don't understand the difference between good and evil. But I have seen them both up close and know that good is beautiful and evil is ugly. And I understand that the person who chooses evil is just like me—not outwardly of course—but has the same mind that I have, and he too carries within himself a piece of the divine. He cannot hurt me because he cannot make me choose evil, but nor can I be angry with him or hate him. For we were born to work together, to cooperate like parts of the same body. So it would be wrong for me to harm him, and if I were to turn my back on him instead of helping him, that would be a harmful act."

Yes. There are people who choose evil, and it would be wrong of us to turn our backs on them. But most people have seen good and know that is beautiful, and they do share the set of values I listed earlier and have throughout history. And those values should be the yardstick by which we measure our progress.

So, by such a yardstick, have things gotten better or worse over the last five thousand years? Definitely better. Not for everyone. Not constantly. Not everywhere. Not without setbacks. And not nearly as fast as we would want. But on balance, over time, inch by inch, I think we are becoming just a little bit better every day.

Is that true? Or just something we wish were true? Let's list some of it out and see if it is convincing. Over five thousand years of recorded history, we've decreased many evils: hunger, poverty, infant mortality, death by violence, death by disease, ignorance, illiteracy, slavery, and war. We've also increased many things that are good: life expectancy, self-government, access to education, individual liberty, legal equality, and wealth. In much of the world, we've outlawed child abuse and child labor, along with cruelty to animals. While doing all of that, we were also working to end legal slavery, the legal status of women as chattel, legal discrimination by race, public torture and executions, debtors' prison, and mutilations as punishment. We have created democracy, human rights, presumption of innocence, trial by jury, rule of law, habeas corpus, and due process. We invented the idea of free speech and codified it in many places around the world, along with freedom of the press and freedom of religion. We enshrined our values in documents that work to guarantee liberty and limit tyranny. While all that was happening, still others were inventing ways to prevent disease, ease the hard toils of life, grow more food, and expand access to information.

While critics could argue with any of these, it is hard to refute them all, and they become all the more compelling when you realize just how bad things used to be. We are attuned to the shortcomings of our time, keenly aware of our own moral failings, and it is right that we should be so, for they are the only problems we have any hope of solving. But as bad as things sometimes seem, it is hard for us as moderns to imagine the cruelty that used to be the norm. I don't even know where to begin to describe it: The smallest offenses used to be capital crimes, mutilations were normal punishments, torture was a form of mainstream entertainment in which people were skinned alive for the spectacle of it. A city that lost a war would have all its men killed, all its children sold into slavery, and its women regarded as spoil. Centuries ago in France, there was a popular pastime known as cat burning, where crowds of people cheered as live animals were thrown into a bonfire. And finally, about the same time, when a rebellion against the Kingdom of Hungary by a man named György Dózsa failed, he was made to sit on a red-hot iron throne, hold a red-hot scepter, and his followers, who had been starved beforehand, were made to pull pliers from a fire, rip the flesh off his living body, and eat it.

What changed from that time to ours? Aurelius would say that we gradually have learned to tell the difference between good and evil, and in doing so, we discovered that good was beautiful, and we went toward it. But how did that come about? Through Agora.

Think about Agora's history and the ways it brought about the progress listed above. That history began several thousand years ago when people started moving into cities. These required governments. Governments were tasked with preserving order, and this was done with legal codes that were clear and, generally speaking, enforced. As individuals we began to cease being laws unto ourselves and were no longer the judge, jury, and executioner of our private legal codes. We became part of a community. Individual motivations must have varied, but whether it was for safety or wealth or something else, we saw some benefit to coming together, like our two selfish bees earlier, and we doubled down on what was working. We specialized and through that became less independent and more reliant on each other, and Agora was born.

Commerce grew, strangers ceased being threats within the safety of the collective with its rules and norms, and instead they became customers and business partners, offering more opportunities for wealth. Increased prosperity led to increased leisure time, which led to more education and thus more literacy. This in turn led to the development of the press, with its ability to galvanize opinion, and this spread of knowledge acted as a check on tyranny. Widespread literacy gave rise to science, which displaced human superstitions. The safety of the city allowed different groups to commingle, and first-person experiences dispelled many of the prejudices and biases that were held in more insular times. New ideas were brought in that mixed with old ones, and slowly we realized the world was a more complicated place than we had thought, and that our ideas and beliefs required more nuance. Moral absolutism waned and the vacuum was gradually filled with tolerance. I can go on and on, but this is ground we've already trod.

We still have much work to do, but that shouldn't discourage us. Bees and ants have had fifty million years to get their superorganism right, and we are less than ten thousand years in on ours.

It's tempting to speculate on the reasons that our great-grandchildren will look upon us with shame, which they likely will. Our descendants will measure us using the same yardstick that we are using, and they will find us lacking. From their vantage point, they may judge us because we built prisons in which to raise animals to eat, who lived short, miserable lives, and how we tried to keep it all out of sight so we didn't have to think about it. Or it might be because we built prisons to hide away the problem people and keep them out of sight for the same reason. Or, it might be that we slept in soft beds with clear consciences in a world with a billion hungry people who could have been fed with little sacrifice. Or how we built weapons that could only be aimed at one possible target: each other. Or, how we made war a for-profit business.

I don't know which of these it will be. Maybe all of them, or maybe something we are completely blind to right now. But I hope they will regard us as monsters, because if they do, just imagine how much progress that means Agora will have made. They will be much better than us at telling the difference between good and evil, and, if all goes well, their great-grandchildren will regard them as monsters, too.

But enough on history. How does Agora help us navigate our lives today? Let's look.

THE PRESENT

CARL SAGAN ONCE OFFERED a solution to the Fermi paradox, that is, why we don't see signs of aliens everywhere if the universe is teeming with life. He said that maybe upon developing the kind of modern technology that we have, a species has about a century—that's a guess, of course—to either rein in their destructive behavior and live on for billions of years or blow themselves up.

It is the vanity of every age to think that they live at some great turning point in history. That said, we actually do live at a great turning point in history. If Sagan is correct, humanity is currently in that narrow window of time when we have the technology to destroy ourselves but may not yet have the wisdom to know how to avoid doing so.

So how do we do it? How do we keep from self-destructing before we make it to the Billion Year Club?

In 1948, English astronomer Fred Hoyle predicted that "once a photograph of the Earth, taken from the outside, is available, a new idea as powerful as any in history will be let loose." He was right. In the decades that followed, we got increasingly better views of the whole planet "from the outside," each one more perspective altering than the last. Then, once people began going into space and seeing it firsthand, they found themselves changed by the experience. Author Frank White coined the term "the Overview Effect," and said that many returning astronauts had experienced changes in how they saw the world. These changes included, per White, "a feeling of awe, a profound under-standing of the interconnection of all life, and a renewed sense of responsibility for taking care of the environment."

Sagan, too, felt that a perspective shift was necessary for humanity to save itself from destruction. In that spirit, he lobbied NASA to signal the distant Voyager 1 probe to turn its camera back toward Earth and take a photo, and they obliged. From such a vast distance, the earth was only a pale blue dot, all alone against a giant black canopy of empty space. The photo, named "the Pale Blue Dot," has no scientific value but enormous emotional value, because from that vantage point it shows that we are unquestionably one people who will share a single fate, and that what we view as our epic conflicts are really just petty squabbles. As Sagan put it, "There is perhaps no better demonstration of the folly of human conceits than this distant image of our tiny world. To me, it underscores our responsibility to deal more kindly with one another, and to preserve and cherish the pale blue dot, the only home we've ever known."

We've probed the limits of reductionism multiple times throughout this book—that how trying to understand something by looking at its parts can make you miss the whole. Sometimes you do have to take a step back to see a thing in its entirety to understand it. The Overview Effect is an example of this. Those astronauts who were moved by seeing Earth from space had seen the very same Earth their entire lives, just from ground level. But seeing it all, in one image, out of one window, in all its majesty and glory, was somehow different.

I think of Agora like that. I've spent my entire life looking at its parts, that is, the people I came in contact with, but I had never perceived them as a whole. Standing back and seeing one single creature, Agora, has given me a new view of humanity.

We are timid creatures, and understandably so. In the distant past, that paid a whole lot better than being bold ones. If we saw a large shape off in the distance on a foggy morning, there were two things we could do. Panic and run off, fearing that it was a bear, or stand fast, optimistically assuming it was just a large rock. It was almost always a rock, of course. That said, every now and then, it was a bear, and when it was, it was only the optimists—along with their opti-mistic genes—who got eaten. The effects of that culling are still with us today.

Our fearful nature expresses itself as an instinctual distrust of things that could be threats to us, and at the top of that list are other humans. We lived so long in small bands comprised solely of our relatives, who by and large looked like us, that we got used to that. Today, we are still just as tribal. The passion with which people identify with certain groups bears testimony to this. I'm not talking about nations and religions here, although that is obviously true. I mean the intensity with which people identify with a political party or a sports team. In a world where riots break out at sporting events, what hope do we have to get along with people who are unlike us in much deeper ways?

Sure, there are two hundred human universals—those behaviors that are common to all human societies—but that's not what evolution has bred us to notice.

What's the solution to this problem? Should we appeal to people's nobler instincts and say, "Why can't we all just get along?" That hasn't worked so far. It's been shopped around by a lot of people for a long time, and folks aren't lining up around the block wanting more of it.

Perhaps Agora might provide a path out of this dilemma. Understanding how Agora operates means understanding that not only do we share common interests, but that we are part of the same creature, the same literal animal. The left hand doesn't have to admire the right ear—it doesn't even have to like it—but it can come to realize that the right ear is part of the same creature as it is, and if that creature dies, well, all the rest of the parts, including the left hand, die as well. There is no possibility for one of them to win and the other to lose.

My hope is that Agora will reframe how we think about each other. It doesn't gloss over our differences; rather it connects us with the simple truth that, like it or not, we are all going to share the exact same fate.

Ahh . . . but what fate? That's the question, right? And that is also the final chapter in our long journey together.

THE FUTURE

MAYBE YOU'VE HEARD ABOUT a 1972 plane crash in the Andes. On this charter flight were forty-five passengers and crew, including eighteen members of the Old Christians Club rugby union, a team based in Uruguay. Because of an error by the inexperienced copilot, the plane began to descend while still fifty miles from the airport. Given that they were over a mountain range at the time, things went poorly. The plane crashed, shearing off both front wings as well as the back of the fuselage. Several people died on impact, and several more would die in the coming weeks from their injuries and the incredibly harsh conditions, which included more than one blizzard. Eleven days into the ordeal, the captain of the rugby team, twenty-two-year-old Nando Parrado, listening to their only working radio, learned that the search for them had been called off. They had been given up for dead. He ran to tell the others, saying, "There's some good news! We just heard on the radio. They've called off the search." The others responded, "Why in the world is that good news?" Nando replied, "Because it means that we're going to get out of here on our own."

When I first read this account, I wondered if it was true. Did Nando *really* say that? And if so, did he believe it? A testament to the amazing world we live in is that within five minutes of asking myself these questions, I found his email address and put them to the man himself. The next morning, I had my answers. The quote was accurate, and on the question of why he said it, he wrote to me, "Given the conditions and fear of the unknown future, I could not come with something better..! Horrible news, that I tried to minimize . . ."

I understand that. He was rallying the troops. But what he evidently didn't understand at the time was that the news they got on the radio actually *was* good news, because it made them stop looking to the sky for their salvation, and once they did that, they had only one other place that they could look for it: within themselves. Once they realized there would be no rescue plane to save them, then they could get to work on the real task at hand: saving themselves. So whether Nando was sincere or not doesn't matter; what matters is that he was right. Those rugby players did just that: they saved themselves.

In the previous chapter, I quoted from Carl Sagan's commentary on the Pale Blue Dot photo. He closed it with, "In our obscurity, in all this vastness, there is no hint that help will come from elsewhere to save us from ourselves. It is up to us."

Sagan's message to us is the same one that Nando heard on his radio. It tells us in no uncertain terms that there is no rescue plane coming to save the human race. We are totally on our own. And you know what? That's good news. It means that we are going to save ourselves.

Today, our world faces seemingly insurmountable challenges. Our situation is grim; our lives hang in the balance. Many have lost hope and have given up. But the message of this book is that though our challenges are great, we are more than up to the task of overcoming them. Or, as Nando might put it, "we're going to get out of this on our own."

This is not rose-colored optimism nor a "We can do it" pep talk. I am not trying to rally anyone. I base this prediction on facts; and, as we reach the end of our journey together, I would like to share my reasoning with you.

Think for a moment just how fragile humanity used to be, how precarious our situation was. Let's go back just ten thousand short years and assess that world and our place in it. At that time, there were just a few million of us, scattered around the globe in tiny groups that had almost no contact with each other. There was no science, no writing, no metallurgy, hardly any technology to amplify our abilities. There weren't medicines beyond a few herbs, no vaccines, no cures, nor even any idea what disease really was. Each person was just a mosquito bite away from malaria, a scratch away from sepsis, and one unlucky injury away from a slow and painful death. Women commonly died in childbirth, children even more commonly died in their first year, and men fared little better, most dying young as well. We lived in the elements, and were hunted by creatures stronger, faster, and more ferocious than us. Every day was a struggle to survive, every night a terror of sounds and shapes. We had no cities, no walls to protect us, no governments, no law codes, no books, no accumulated knowledge but the little we could remember and pass down. In our ignorance, superstition must have ruled our lives, and the world must have been a frightening place. Lightning and thunder would have been terrifying, as would wind and storms. How confusing life would have been, how capricious death must have seemed.

And yet we made it. We got through it.

Now consider us today in all our glory, in our power and might. I won't go through that litany—you can look around the world and see us numbered in the billions, and marvel at our mastery of the planet, our monumental accomplishments, our inexhaustible knowledge, and our magnificent technical achievements. You already know that we live in a world where breakthroughs in science come so quickly no one can even know the tiniest fraction of them, and how a thousand things that would have seemed like miracles to our forebears are now parts of our daily lives.

Am I exaggerating even the tiniest bit?

Agora is staggeringly powerful and grows more so by the moment. We can now face anything that fate can throw at us, and we can overcome it . . . but only if we are willing to work together to do it. That's the trick, that's the power of a superorganism. But what exactly does that mean on a practical level?

When each of, as individuals, stops and considers our own place in Agora, it is easy to feel tiny and inconsequential. After all, there are billions of us who make up Agora, many of whom are renowned for their accomplishments, for what they have built or discovered. There are all those famous people whose names we all know who seem to be the ones doing the important work while the rest of us are living out our regular lives, caught up in the daily struggle of just getting by.

But it would be a mistake to think of Agora that way. That's just not how superorganisms work. What they achieve is done through the interactions of their parts, not because of some genius bee or overachieving ant. The power people are not the backbone of Agora; they aren't any more important than anyone else. I know that might not seem to make sense, or maybe it seems like I am being patronizing, but it is completely true. Are the biggest gears the most important parts of the clock? No. The clock keeps time because of all of the parts, big and small.

This is what really makes Agora work: There's a kindergarten teacher who encourages a child, who grows up to paint a mural that brightens someone's day as they drive past it, who, in turn drops back a bit in traffic to let the car in front of them change lanes, whose occupant thereby arrives to work in a slightly better mood, and is thus able to comfort a coworker who is struggling at home, and on, and on and on, year by year. All of these sorts of small acts are what sustain Agora, not the noisy people doing the glitzy things. By itself, any one of these acts of kindness is tiny, and is probably done reflexively, without thought, but when they occur by the billions every day, that's what brings Agora to life, that's what makes it strong.

"But wait," you may be thinking. "Let's not be overly sentimental here. The sequences of events that our good actions set in motion can just as easily result in harm, right? We can't be sure how it will all play out because we can't anticipate even one step ahead, let alone a hundred."

Shortly after World War II, Dr. Jerome Motto, a psychologist who specialized in the study of suicide, visited the apartment of a man who had just killed himself by jumping off the Golden Gate Bridge. The deceased had been single, in his thirties, and lived alone. But he had left a note on his desk, addressed to no one in particular, just for whoever happened to find it. It read:

"I'm going to walk to the bridge. If one person smiles at me on the way, I will not jump."

Superorganisms can only thrive when the parts work in harmony, cooperating with each other, assisting one another. In humans, this behavior comes not from instincts but emotions, specifically love, compassion, and empathy, all of which affirm life. How do you think we made it through those hard times ten thousand years ago? Through ruthless utilitarianism? A fierce survival-of-the-fittest mentality? A dog-eat-dog view of life? No. Just the opposite: through love, compassion, and empathy. This is not romanticism on my part. Archeological remains from that time show many people who clearly had sustained injuries that required constant care, and yet we can tell by the way their bones grew afterward that they lived on for years. We see repeated examples of adults from that era who had been born with debilitating conditions that would have prevented them from being able to contribute to the group at all, and yet they lived into adulthood. We see the skeletons of elderly people who must have been unable to walk for years before their death, but they were not put out to die. We made it through those hard times because we were kind to each other, because we helped each other.

So, OK, sure. We can't know, *for certain*, that our kind acts will result in good ends, but we can be confident that they almost always will. A loving, selfless act doesn't ripple through the world and somehow promote hate and division. It simply doesn't. Someone might be able to construct a carefully contrived scenario involving baby Hitler or something, but these are just arguments used by people who want to justify their own unkindness, or, at best, their own apathy. In truth, kind acts do promote life.

So back to those great challenges that we as a species face. We will overcome them not through some scientific breakthrough or due to a charismatic leader but by countless acts of life-affirming kindness, which in aggregate vitalize Agora, who is more than up to any task. Thus our collective fate depends on our innate goodness, which we can be 100 percent certain exists, because if it didn't, we never would have made it this far. We never would have gotten out of that world of ten thousand years ago.

Isn't it a wonderful truth that the emotions that come naturally to a child—kindness and empathy—turn out to be the most powerful forces in the universe? For they promote life, and collectively life becomes Agora, which exists to protect and spread more life.

If you've ever felt like you aren't living up to your potential, or that you should be doing more with your life, I suggest you move past all that. Remember that no part of a superorganism can comprehend the whole. Agora is far too big for any single person to shape or move. Instead, it can only be influenced by small acts of kindness done in great numbers. So, put no heavier burden on yourself other than to be as kind as you can be, and try every day to be a slightly better person than you were yesterday. That really is all it takes to build utopia.

We can build a better world, and then build an even better one, and then one even better than that. Until one morning we wake up in a world we cannot imagine any better. And that is the world we will take to the stars, spreading it to a billion planets, populating each with a billion people, each of whom is empowered to live their best possible life.

We really can do this. How do I know?

Because we are Agora.

ACKNOWLEDGMENTS

AHH, THE ACKNOWLEDGMENTS. THAT'S the part of the book analogous to the gigantic credit roll at the end of a movie. You know, the ones with the shout-outs to the third best boy and the caterer on the Santa Fe shoot.

My prior books have had such credit rolls, and as I sat down to write this, I realized that if I took that same approach, it would almost just be a copy and paste from my last book, and that didn't feel quite right to me. What's the point of that? So, I am doing this one a bit differently. I want to acknowledge only one person, my editor, Alexa Stevenson. We live a thousand miles from each other, and have never met, and probably never will. In fact, we've had less than a dozen calls. But her notes on my last book were 30,000 words long. That's on a book that itself was just 80,000 words. This one wasn't much different, with her notes weighing in at 25,000 thousand words. And the humbling thing I must admit is that she was right about everything. Everything. Every. Single. Thing.

Without her, my last book and this one would have lacked the organization, clarity, structure, and consistency needed to convey the ideas I wanted to get across. So, she has saved me from myself twice, and I hope will do so again.

Thank you, Alexa.

APPENDIX

MEDITATIONS, BY MARCUS AURELIUS, Book II, Paragraph I, from George Long's 1910 translation:

Begin the morning by saying to thyself, I shall meet with the busy-body, the ungrateful, arrogant, deceitful, envious, unsocial. All these things happen to them by reason of their ignorance of what is good and evil. But I who have seen the nature of the good that it is beautiful, and of the bad that it is ugly, and the nature of him who does wrong, that it is akin to me, not [only] of the same blood or seed, but that it participates in [the same] intelligence and [the same] portion of the divinity, I can neither be injured by any of them, for no one can fix on me what is ugly, nor can I be angry with my kinsman, nor hate him. For we are made for co-operation, like feet, like hands, like eyelids, like the rows of the upper and lower teeth. To act against one another then is contrary to nature; and it is acting against one another to be vexed and to turn away.

INDEX

ABOUT THE AUTHOR

Byron Reese is an Austin-based entrepreneur with a quarter-century of experience building and running technology companies. He is a recognized authority on AI and holds a number of technology patents. In addition, he is a futurist with a strong conviction that technology will help bring about a new golden age of humanity. He gives talks around the world about how technology is changing work, education, and culture. He is the author of four books on technology, his most recent was described by the New York Times as "entertaining and engaging."

Bloomberg Businessweek credits Byron with having "quietly pioneered a new breed of media company." The Financial Times reported that he "is typical of the new wave of internet entrepreneurs out to turn the economics of the media industry on its head." Byron and his work have been featured in hundreds of news outlets, including the New York Times, Washington Post, Entrepreneur Magazine, USA Today, Reader's Digest, NPR, and the LA Times Magazine.